Gwynne's Evolution or C~~   ?

Formerly a success~~~~~~~~~~~~~~~~~~s, for many years, been teachin~~~~~~~~~~~~~~~~~~'bout every sort of pupil in just ab~~~~~~~~~~~~~~~~lish, Latin, Greek, French, Germ~~~~~~~~~~~~~~~~~ical philosophy, natural medicine, the ~~~~~~~~~~~~ and 'How to start up and run your own business ~~~ lecture halls, large classrooms, small classrooms and homes – to pupils from three years old to over seventy – of many different nationalities and in several different countries – and, since 2007, 'face-to-face' over the Internet.

His teaching methods are very much in accordance with the traditional, common-sense ones, refined over the centuries, that were used almost everywhere until they were abolished worldwide in the 1960s and subsequently.

His teaching has been considered sufficiently remarkable – both in its unusualness in today's world and in its genuinely speedy effectiveness – to have featured in newspaper and magazine articles and on television and radio programmes.

Previously published books by him are:

Gwynne's Grammar: The Ultimate Introduction to Grammar and the Writing of Good English. (Ebury Press, London, 2013.)

Gwynne's Grammar: The Ultimate Introduction to Grammar and the Writing of Good English. (Alfred A. Knopf, New York, 2013.) An expanded version of the Ebury Press edition.

Gwynne's Latin: The Ultimate Introduction to Latin Including the Latin in Everyday English. (Ebury Press, London, 2013.)

Gwynne's Kings and Queens: The Indispensable History of England and Her Monarchs. (Ebury Press, London, May 2018.)

The Truth about Rodrigo Borgia, Pope Alexander VI (Lulu.com, July 2008)

GWYNNE'S
EVOLUTION or CREATION?

An All-Important Subject Investigated

N M Gwynne

St Edward's Press Ltd

Formal Notes

Gwynne's Evolution or Creation?

First published in 2020 by
St Edward's Press Ltd
20 Barra Close
Highworth
Swindon
Wilts
SN67HX

1st edition

All rights reserved. No part of this publication may be reproduced, stored in a retrieval system, or transmitted, in any form, or by any means, electronic, mechanical, photocopying, recording or otherwise, without the prior permission of the publisher.

N. M. Gwynne has asserted his right to be identified as the author of this Work in accordance with the Copyright, Designs and Patents Act 1988.

While every effort has been made to ensure that this publication by St Edward's Press Ltd provides accurate information, neither the publisher, author, retailer nor any other supplier shall be liable to any person or entity with respect to any loss or damage caused by, or alleged to be caused by the information contained in, or omitted from, this publication.

ISBN 978-1-909650-10-7

© N M Gwynne

Cover design by 659design, Highworth
Printed by Ashford Ltd, Gosport.

GWYNNE'S EVOLUTION OR CREATION?

CONTENTS

PART I. Introductory — 1

PART II. On behalf of the theory of evolution. — 23
1. The theory of evolution introduced. — 25
2. The theory of evolution as such. — 29
3. The fossil evidence. — 36
4. Speciation. — 39
5. Vestiges and embryology. — 45
6. "Biogeography". — 50
7. Natural selection. — 53
8. Evolution's most impressive achievement? — 55
9. Summing up with Professor Coyne. — 59
10. Is there anything to add? — 60

PART III. Are there, however, any problems with the theory of evolution? — 67
1. Introductory. — 69
2. Can one species change into another species? — 71
3. Laws of nature. — 80
4. Evidence of design in the operations of nature: the eye; the digestive system; feathers; wing-movement in insects; the honey-bee; instinct; the ratios of the sun, the moon and the earth in relation to each other. — 88
5. Evidence from fossils. — 107
6. More on fossils. — 110
7. Dating of the universe and its contents, by means of, for instance carbon-14, fossils, and radio-metric dating. — 117
8. On the relevance of the earth's magnetic field. — 122

9.	From hydrogen to humans.	124
10.	Then why was the theory of evolution introduced? And why has it survived?	129
11.	Darwin's problems with his own theory.	131
12.	Summing up.	135
13.	Interestingly, however…	138
14.	Post Script. Some scientists on science as it is today.	146

Bibliographical	149
Acknowledgements	155

PART I
Introductory.

Introductory

This book's purpose is to set out a strictly scientific treatment of its subject. That is to say, its purpose is to assess, with due care, the only two widely held explanations of how we human beings came to be where we are:

1. *Evolution* – the theory first promoted extensively by Charles Darwin in the mid-19th-century. This explanation is of course widely accepted today as being the correct one.

2. *Creation* – the creation, planned and systematic, of the universe and its contents by a "Being" capable of bringing this about. This explanation is generally ridiculed today and in some circumstances is even a forbidden topic for discussion.

What this book therefore sets out to be is a *scientific* book of science.

* * * * *

It is as well at the outset to establish first of all exactly what is meant by the term "science". Although this is perhaps an unusual approach in books dealing with science, I submit that it is indispensable in this case. What value, after all, can any discussion of such a topic have if there is not complete clarity on exactly *what* it is that is being discussed? All the more is this so, moreover, in our present day when virtually no one – and this is very much true in the world of today's scientists – knows the meaning of the word science, as it is represented by the primary definition of it in traditional English dictionaries such as many of the dictionaries published by the Oxford University Press.

I recognise that it may not be to every reader's taste to have to work through an introduction that may seem to be a digression. Accordingly, I invite those who wish to get straight into the arguments represented by this book's title to go directly to chapter 1 of PART II, on page 23. I urge those who do this, however, to return here at a later stage, because these pages are in fact as important as any of the rest of the book, for reasons which will be obvious when they are read.

*   *   *   *   *

Addressing now the word "science" and exactly what it means...

Let us take as a starting point the fact that acquiring knowledge about a subject consists of finding out at least *some* of what is true about it.

Already we have a difficulty. Go today to a philosophy course in a school or university almost anywhere in the Western world and it is likely that you will be assured that there is *no such thing* as objective, attainable truth. *All* truth, you may be told, is *relative*. For each of us, *our* truth is what is true for *us*, and *only* for us, individually.

Most children who have reached the age of reason, at around six or seven years old, can at once see that this version of the meaning of truth is simply – dare I say it? – not true, and that what is true is true for *everybody*. If, for instance, a boy trips over and cuts himself, this is a truth not just for him personally, but for anyone who has seen it happen. He and every witness of the event knows that it is *true* that it took place, and that it is *not* just his own personal truth. The great majority of highly qualified philosophers, academics and scientists, however, will *continue*, somehow, to deny that that there is such a thing as *objective* truth. Interestingly, children can quite often see reality where the most highly qualified experts are unable to.

An important reason for this strange phenomenon of experts-versus-children in relation to knowledge is that people who have reached adulthood are all too often victims of bad mental habits that – from such influences as their parents, their teachers and their general environment – they have acquired over the years. A typical bad habit of this kind is for us to be guided in our beliefs, even *dictated to* in our beliefs, by what we *prefer* to believe, rather than by reliable evidence. This is not surprising if we realise that facing up to truth can sometimes be genuinely painful. As the 19th century German philosopher Friedrich Nietzsche put it:

> It is no requirement of truth that it should make one feel good.

The notably influential 17th-century statesman and author Lord Halifax was even more emphatic:

> Nothing has an uglier look to us than reason when it is not on our side.

Let no one be mistaken: this psychological factor is real, and, in its effects, sometimes even frighteningly real. It is also relevant to the effect that the subject-matter of this book will have on this book's readers. Sadly, there can be no doubt that all too many of those who are confronted with what is in these pages will be primarily affected in their judgement, *not* by the *reasoning and evidence* supporting one side or the other, as of course they ought to be, but, rather, by their *present beliefs or convictions*, in effect by their mental *habits*. For whatever reason in each case, evolutionists will tend to want to remain evolutionists, and anti-evolutionists – or whatever they prefer to call themselves – will tend to want to remain anti-evolutionists.

Please, good reader, do not be one of those people. *Please* make a firm resolve to be guided by reason and adequate evidence, *only* by reason and adequate evidence, and by nothing else.

* * * * *

Perhaps the first thing worth saying about science is that, as has been recognised for most of recorded history, *real* science, in what has always been that word's primary meaning, is different from what almost everyone today supposes it to be. *Real* science is knowledge systematically acquired on *any subject whatever* that has reality attached to it. This until recently was the first definition of science given in the *Concise Oxford Dictionary*:

> Systematic and formulated knowledge (moral, political, natural, etc.).

Fundamentally, therefore, science is what we know, or can know, about *anything*. Today the word "science" represents what was formerly called "*natural* science". The areas embraced by natural science were physics, chemistry, biology, astronomy and so on.

Here, now, are two important facts relating to *all* branches of science – not merely natural science – arising from the very nature of science under the primary definition of it just given.

The first fact is that, for any branch of science to be *true* science, it must be true *in every detail*, even the smallest details. If any detail fails this test, the *whole* of that branch of science is rendered unreliable. It is actually not even helpful for an incorrect detail to be *close* to the truth; and in fact the very opposite can be the case, because error that is close to truth is, as one would expect, more likely to deceive people than error that is further removed from truth.

The second of the two important facts under this heading is that no single branch of science can *validly* be at *any* point in contradiction to any other branch of science. If such a contradiction exists, either *one* of the two branches of science is wrong, or *both* of them are wrong. Something that is genuinely true can *never* be validly contradicted by something *else* that is genuinely true – not even if the two are in completely different areas of science.

This can easily be shown to be true with an example relating to the particular theory that we shall be concentrating on, that of evolution.

According to this theory:

(a) the universe is some billions of years old; and

(b) during the vast period of time that it has existed, all living things on Earth gradually evolved into their present forms.

By contrast, those – such as the majority of followers of one or other of the countless different forms of Christianity, and of the followers of Judaism, and of the followers of Islam – who believe in principle that the Old Testament of the Bible is true in what it records or purportedly records, necessarily believe the universe to be only a few thousand years old, with all living things basically the same as they were at the beginning.

Clearly those two contrasting beliefs cannot *both* of them be true, from which it follows:

either that one of them is right,

or that the other of them is right,

or that they are both wrong.

And the same applies whenever any other area of science is in definite contradiction with another area of science.

* * * * *

In relation to all that has been said so far, there are two crucially important questions:

First, how -- by what means -- can we establish what is true, definitely true, on any particular matter?

Secondly, how can we be *certain* that what we *believe* to be true really *is* true?

The answer to both those questions in combination, even though it is straightforward and virtually self-evident once stated, is, shockingly, an answer that is now taught to students almost nowhere in the world, whether in philosophy classes or elsewhere in education.

There are in fact three ways of arriving at certainty, to be applied in accordance with the nature of whatever subject that is under examination.

The first and most obvious way is that of *sense experience*: the experience of one or more of our five senses of sight, hearing, touch, taste and smell. Any of these senses can be completely relied upon if properly used.

It is true that we can sometimes be misled in what we think our senses are telling us, most obviously by professional conjurors. In most situations, however, there is no room for doubt – for instance, our sense of touch if we shake hands and our sense of sight enabling us to distinguish between daytime and nighttime.

The second way is by *logic* – that is to say, by using the process technically known as the syllogism.

The syllogism works as follows:

1. We start with a definite *general* truth, such as that boiling water is painful to the touch. In this context, this is known as "the major premise".

2. We then take a *particular* truth, such as that you are at present holding a cup of boiling water if indeed you are. This is known as "the minor premise".

3. Combining these two facts leads to a definite *conclusion*, known as "the conclusion" – in this instance the conclusion that, if you were to touch the water in that cup, you would experience pain.

That may appear complicated because of the technical terms "syllogism", "premise" and "conclusion" that I have used, but it is completely straightforward.

For instance:

(a) if it be known, as of course it *is* known, that direct contact with boiling water is painful to us, which is *the major premise*,

and (b) we put one of our fingers into a cup of boiling water, which is *the minor premise*,

we *know* in advance that doing this will cause us pain, which is *the conclusion*.

Equally, if we know, as the *major premise*, that a male dog mating with a bitch will result in puppies of the dog species if there is any result at all, we can know in any *particular* instance of such a mating, the *minor premise*, that the birth of puppies is what will happen, rather than the birth of some other animal: *the conclusion*.

The third way, and the only other way, of arriving at certain knowledge is faith, properly defined.

This assertion may surprise many readers. Is not faith a sort of blind trust in what we are told in religious matters?

Those who think that are in learned company. For instance, this is what an evolutionary biologist some of whose work we shall be looking at later on, Professor Jerry A. Coyne, Professor Emeritus in the Department of Ecology and Evolution in Chicago University and a best-selling author on the theory of evolution in particular, emphatically says in the Preface of a book by him, *Faith Versus Science*, that was published in 2015:

> **Faith may be a gift in religion, but in science it's poison, for faith is no way to find truth.**

Many other people well qualified to pronounce on such subjects would agree with him, of course; but the fact that an assertion is authoritatively made and widely believed does not make it certain that that assertion is true; and this particular assertion is in fact *drastically* incorrect.

For a start, it is not in accordance with the first definition of faith in the standard Oxford English dictionaries, which is "belief founded on authority" or some equivalent of that.

Does that not amount to much the same thing? No, faith properly considered -- *true* faith -- is belief founded on authority that we have *adequate reason* to trust as being *undoubtedly reliable*. For instance, it is by faith, and *only* by faith, that we can know such things as:

> the date of our birth;

> that a country that we have never visited exists;

> that Queen Victoria was the reigning queen of Britain and its empire during much of the nineteenth century;

> that there was a world war in 1914-18;

> and so on.

The reality is that *most of what we know*, and know for certain, we know by faith and *only* by faith; and it is certainly remarkable that Professor Coyne, the author of the book just quoted from, *Faith Versus Science*, should mis-define one of the three words of his book's actual title.

But… is not *religious* faith true faith as well, but simply a different *form* of faith?

Not so: religious faith, as with any other form of faith, can *only* be faith founded on adequate authority; and any belief *not* so founded cannot be made into real faith simply by *calling* it faith. Here is the distinguished religious correspondent of the London *Daily Telegraph* Christopher Howse, confirming this in an article dealing with religious faith in the *Daily Telegraph* of 6th June 2019:

> **Faith is not belief based on an absence of evidence, but belief based on the testimony of someone reliable…**

Returning now to what exactly is meant by the word science: to deduce by logical steps – that is to say, by clearly *valid* logical steps

– that, for instance, God exists and created the universe and its contents, if such deducing can be done, is just as much to focus one's attention *scientifically* as is to study astronomy or physics or the theory of evolution promoted by Darwin.

*   *   *   *   *

It is as well to recognise that *true* religious faith – on the assumption, *only* provisional at this stage, that there is such a thing – can at most be held by believers in *only one* religion. This is because

(a) if two religions are in any way in contradiction to each other, they cannot *both* of them be true, as we have seen,

from which it follows

(b) that any supposed authority backing up any untrue religion must be a flawed authority.

It is certainly *logically* possible that all religions are *false*, and that everyone trusting in them is deceived as to his or her reasons for believing in them, although I am not suggesting that that is so in reality. What is *not* possible, however, is that *more than one religion* can be *true*.

*   *   *   *   *

Another relevant fact that is of great importance in this context, and is a further development of some of what has been said above, is that truth is not decided by majority vote, nor even by unanimous vote. This was highlighted in the following celebrated statement attributed to Abraham Lincoln, one of the best-known Presidents of the United States:

> **You can fool all the people some of the time and some of the people all the time, but you cannot fool all the people all the time.**

Yes, according to Abraham Lincoln, it really *is* possible to fool *some* people perpetually.

Important as well: truth is not decided by *experts* on any particular subject even if they are unanimous in their positions on it, and this includes experts in the scientific world.

That last statement, assuming that it can be shown to be true, is indeed of great importance, given the general submissiveness of the public to such experts. I am therefore pleased to be able to offer as solid a proof in its support as could reasonably be asked for: the experience of the leading expert on the theory of relativity in his day, the late Professor Herbert Dingle, D. Sc., D.I.C., A.R.C.S., Emeritus Professor of History and Philosophy of Science, University College, London, a member of the Royal Society, and a former President of the Royal Astronomical Society – thus both a leading scientist and a professional philosopher.

Might there be readers who feel inclined to wonderment at my embarking, however briefly, on a discussion of *relativity theory* in a book about evolution and the alternative to it? To any such readers I have a ready answer. At least in Great Britain, and to a considerable extent elsewhere in the world as well, *all* the natural sciences today – the earlier-mentioned physics, chemistry, biology and astronomy, and certainly including evolution – and how they are investigated have been looked after, preserved and developed by the organised scientific community that has for centuries been headed by the Royal Society. That society was founded back in 1660 in the reign of King Charles II, with its "fundamental purpose" being, in the words of its present-day manifesto, "to recognise, promote, and support excellence in science and to encourage the development and use of science for the benefit of humanity." The Royal Society has for long enjoyed unmatched prestige all over the world and has wielded great influence.

All the various sciences are treated by the Royal Society in accordance with the same time-honoured principles. How any *particular* science is treated by the Society can therefore safely be taken as fully representative of how all the other sciences are treated by it.

And the reason for my bringing relativity theory into the arena?

As it happens, an event of a few decades ago in the context of the branch of science known as the theory of relativity gives a notable and obviously pertinent example of how the Royal Society handles matters if and when a member comes up with a finding that is in contradiction to what is generally accepted as the correct science in any particular scientific subject. In other words, a *particular* experience of Professor Dingle in the context of his specialist subject, relativity theory, also has an important *general* relevance in the context of science; and "general relevance" of course includes relevance to the scientific topic addressed in this book. Putting the story before my readers will therefore have two important benefits even though the story has no direct connection with evolution theory:

First, it will make clear what the scientific community is prepared to do, *and* is capable of doing *worldwide*, whenever there is a threat to an existing scientific orthodoxy.

Secondly, it will illustrate and underline the crucially important philosophical scientific principle stated above: that truth is not decided by *experts* on any particular subject even if they are unanimous in their positions on it, not even by experts in the scientific world.

Here, therefore, is the story relating to Professor Dingle that I offer as relevant to our topic discussed in this book even though it is not *directly* related to it. Those readers who are anxious to start exploring the theory of evolution as soon as possible are welcome to skip to the last page or so of this PART I, although it is very much my hope that they would return here at some point, because what now follows really is relevant to much of what is in the rest of this book.

The qualifications of Professor Herbert Dingle for being considered an authority on relativity theory literally could not be more credible or convincing. Here is what he was able to say on this at one point in his third and the most important book on relativity, his *Science at the Crossroads* (published by Martin Brian & O'Keeffe, London, 1972). On page 105, after indicating some embarrassment at the seemingly ungentlemanly immodesty that he was displaying, but having decided that he was compelled to present the situation to his readers faithfully and completely, he went as far as to say:

> **To the best of my knowledge there is no one now living who can give objective evidence that he is more competent in the subject than I am.**

Well, either that claim is justified or it is not, irrespective of any modesty or immodesty involved in making it.

In the first place, it has never been publicly contradicted, even though many establishment scientists must have ached to do so.

In the second place, and much more important, the evidence in support of it is undeniably impressive, indeed overwhelming.

Because of the importance of the *real* status of experts as authorities that we can and should put our trust in, that we must keep constantly in our minds, it is well worth looking at this evidence.

In 1922, three years after the theory of relativity first attracted the attention of the public, Professor Dingle wrote *Relativity for All*, one of the very first textbooks on the subject. During the following fifty years after its publication he continued an intensive study of the theory, amongst other things discussing it with all the physicists whose names are best known in connection with it: people such as Einstein, Eddington, Tolman, Whittaker, Schroedinger, Born and Bridgman.

A second book by him, *The Special Theory of Relativity*, remained for many years a standard work on the subject, routinely used in English and American universities. Furthermore, one of the two articles on the subject of relativity in *Encyclopaedia Britannica* was written by him; and when the founder of relativity theory, Albert Einstein, died, Dingle was chosen by BBC Television to broadcast a tribute to him.

Against this background, let us look briefly at two passages in particular in his published writings. I ask my readers to read them patiently and carefully, because what emerges from them is fundamental to some of what will be following in the rest of this book.

The first is from the Preface in the last edition -- which appeared in 1961 -- of the book by him just referred to, his *The Special Theory of Relativity*:

> Since this book was written, reasons have appeared, which to me are conclusive, for believing that the theory is no longer tenable. Though this is not yet generally accepted, it has not been questioned that, so far as experimental evidence goes, an alternative theory is equally possible. This is quite a different situation from that existing previously, when the theory seemed the only possible interpretation of the facts of experiment. My first impulse was to withdraw the book from circulation, but on second thoughts it seemed more fitting to issue it with an explanation of the present position in relation to the presentation of the theory given here, and this I now proceed to give.

What had happened was that in 1959 Dingle had suddenly woken up to an impossible internal contradiction in the theory, one indeed that was so obvious that it would not be beyond the ability of a schoolboy or schoolgirl to recognise it immediately once it was pointed out. This was simply that the theory required *each* of two clocks to work *faster than the other, each* of two twins to age *more slowly than the other, each* of two masses to *be greater than the other, each* of two measuring rods to be *shorter than the other*; and so on. No, I am not exaggerating either the simplicity of Dingle's discovery or the obviousness of the fallacy. Here is what Dingle himself said on page 17 of his *Science at the Crossroads*:

It would naturally be supposed that the point at issue... must be too subtle and profound for the ordinary reader to be expected to understand it. On the contrary, the point at issue is of the most extreme simplicity. According to the theory, if you have two exactly similar clocks, A and B, and one of them is moving with respect to the other, they must work at different rates, i.e. one works more slowly than the other. But the theory *also* requires that you cannot distinguish which clock is the "moving" one; it is equally true to say that A rests while B moves and that B rests while A moves.

The question therefore arises: how does one determine, in accordance with the theory and *consistently* with the theory, which clock works the more slowly?...

Unless this question is answerable, the theory unavoidably requires that A works more slowly than B, *and* that B works more slowly than A, which it requires no super-intelligence to see is impossible. Now, clearly a theory that requires an impossibility cannot be true, and scientific integrity requires, therefore, *either* that the question just posed shall be answered, *or else* that the theory shall be acknowledged to be false.

What do you think of *that*, my gracious readers?

For thirteen years Dingle continued to ask for an answer to his clear and straightforward proof that Einstein's special theory of relativity was false. His request, which of course amounted to a challenge, was met in consistent fashion by all the most distinguished scientists in the world, by the Royal Society, by scientific journals in England and America, and even in the lay press, with a single exception. In Dingle's words on page 15 of his *Science at the Crossroads*, what he asked was...

...ignored, evaded, suppressed and indeed treated in every possible way except that of answering it, by the whole scientific world.

For a brief summary of the results of Dingle's struggles to obtain an answer to his question, which incidentally are recorded in full and enthralling detail in his book, very helpful is this letter by the Rev. W. J. Platt that was published in the London *Times*. It was written as a result of Platt's having read correspondence in the only periodical which did not refuse to handle the subject, the then-weekly (now defunct, as of 1991) *The Listener*. Dingle reproduces it on page 91 of *Science at the Crossroads*.

> In the *Listener* last year there appeared a long correspondence following an article entitled "Definitions and Realities" by Professor H. Dingle, which was published on July 3. In its course, certain alleged facts transpired which, if true, are manifestly of public concern. I have been waiting for some authoritative statement showing either that the assertions were unfounded or that steps were being taken to rectify a dangerous situation. As far as I am aware, none has appeared, and the implications of the matter seem so serious that public interest demands one without delay...
>
> The situation thus disclosed, if the facts are as stated, is alarming. According to Dingle's closing letter (30th October), all that is required to settle the matter is an answer to the question: "What is it, on Einstein's theory, that determines *which* of two clocks, relatively moving uniformly, lags behind the other, as Einstein says?"
>
> Dingle's contention is that, to be true, the theory demands that the clocks must work faster and slower *at the same time*, and that it is therefore untenable. I repeat, Sir, that I make no attempt to judge the issue, but ask, in the public interest, since the foregoing assertions have been published and remain uncontradicted, that an authoritative and conclusive assurance shall be given that scientific integrity continues to exist.

With that background in place, here, now, are some further revealing extracts from the Introduction in *Science at the Crossroads*, beginning on page 15 and including, in the second paragraph, a passage that I have already quoted a few paragraphs back.

This is a book that I have been trying for more than thirteen years to avoid having to write. I have at last been forced to do so because it has become impossible for its purpose to be achieved otherwise, and that purpose is imperative.

I can present the matter most briefly by saying that a proof that Einstein's special theory of relativity is false has been advanced; and has been ignored, evaded, suppressed and, indeed, treated in every possible way except that of answering it, by the whole scientific world. (The world of *physical* science, that is; the theory has no place at present in the biological and psychological sciences.) Since this theory is basic to practically all physical experiments, the consequences if it is false may be immeasurably calamitous. That is why the failure of physical scientists to practise what is generally understood to be their faithfully preserved fundamental ethical principle – the subordination of all theories, however plausible, to the demands of reason and experience – compels its exposure.

It will in all probability immediately strike the reader that the theory of relativity is believed to be so abstruse that only a very select body of specialists can be expected to understand it. In fact this is quite false; the theory itself is very simple, but it has been quite unnecessarily enveloped in a cloak of metaphysical obscurity which has really nothing whatever to do with it. The physical theory itself, indeed, is much simpler than metaphysical theories familiar to most educated non-scientific but interested persons in the 19th century....

Briefly, the great majority of physical scientists, including practically all those who conduct experiments in physics and are best known to the world as leaders of science, when pressed to answer allegedly fatal criticism of the theory, confess *either* that they regard the theory as nonsensical but accept it because the few mathematical specialists in the subject say that they should do so, *or* that they do not pretend to understand the subject at all, but, again, accept the theory as fully established by others and therefore as being a safe basis for their experiments.

The response to criticism made by the comparatively few specialists is either complete silence or a variety of evasions couched in mystical language which succeeds in convincing the experimenters that they are quite right in believing that the theory is too abstruse for their comprehension, and that they may safely trust those who are endowed with the better physical and mathematical talents that enable them to write confidently in such profound terms. What *no one* does is answer the criticism.

I remind readers of how Dingle concluded a passage by him that I quoted earlier:

Now, clearly, a theory that requires an impossibility cannot be true, and scientific integrity requires, therefore, *either* that the question just posed will be answered, *or else* that the theory will be acknowledged to be false. But, as I have said, more than thirteen years of continuous efforts have failed to produce either response...

This, now, is how he followed up that passage:

It is no doubt generally believed that means exist for preventing the occurrence of such a situation as this, and theoretically they do. The Royal Society is a body whose function includes the safeguarding of scientific integrity in all matters, and especially those vital to public welfare in this country; and, accordingly, after great difficulty overcoming the interposed obstacles, the criticism was submitted to it for consideration.

It was rejected on the basis of a report from an anonymous "specialist" that the fallacy invalidating it was too elementary even to be instructive. The "fallacy", however, was not revealed; nor was this simple but crucial question answered; but the customary paragraphs of mystical comments were supplied, and these satisfied the Society that the question was baseless.

A letter to the leading scientific journal, *Nature*, asking, in the public interest and in accordance with the principles of the Society, that the fallacy should be published, was refused publication, on the ground that actions of the Royal Society were not open to question in *Nature*.

An attempt was made to obtain a ruling of the Press Council (one of whose functions is "to keep under review developments likely to restrict the supply of information of public interest and importance") on this refusal by *Nature* – not, be it noted, merely in this instance, but on the general decision of the editor that no action of the Royal Society, whatever its relation to the public interest, was open to questioning in the journal – but the officers of the Council would not allow the enquiry to reach it. Other scientific journals impose a similar veto.

That again is part of the reason why I have been forced to use the medium of the book to acquaint the public with the position in which it stands: a body of scientists, in whose uncontrolled hands the physical safety of the whole community lies, is daily engaged in experiments of the greatest potential danger, based on principles which the experimenters confess they do not understand, and the Press is closed to any criticism, however well informed, of their activities, and to all questionings of their decisions.

It is of no little interest that, despite Dingle's exalted status in the scientific world and the book's being, at least in my judgement, a magnificent piece of literature in its own right, and, amongst much else that makes it highly readable, its telling a genuinely thrilling story, Dingle found it almost impossible to find a publisher for it. He finally did find a small publisher, but neither he nor the publisher could get it reviewed anywhere and the number of copies of it that were sold was minimal. It was only by a considerable and unlikely piece of good fortune that I came across it myself.

Deeply shocking, all the foregoing, I suggest. Moreover – and this is why I have put the saga before my readers in a book not devoted to relativity theory – if such extreme intellectual dishonesty, perpetrated on a massive scale by the leading authorities in science, can occur in the realm of relativity theory, we must clearly consider ourselves to be "under alert" to the realistic possibility that it could happen in *other* scientific disciplines as well.

*   *   *   *   *

In summary, for our particular purposes, of what Dingle has been telling us: we now have objective proof that what might be called "the scientific world", of which England's Royal Society, founded by King Charles II, is the world's most distinguished and admired organisation, *is not truly scientific at all*, and in fact is very much the reverse. Fundamental to its existence is a *determination* – and, as we have seen, an astonishing *ability* – to impose on everyone, including us "ordinary folk", what the leading members of the scientific community are determined to impose on us, as to what we are to believe and what we are to teach our children and so on, and to do so for reasons that they keep concealed.

And to any reader who may think that this picture of the world of science exists only in the arena of relativity theory, I make the suggestion that we should keep in our mind at least the *possibility* that it may apply in other areas of science as well.

What this means in practice is that, in our examination of the topic of whether (a) evolution, and everything necessarily associated with it, is the true explanation of the existence of our universe and its contents, or (b) carefully-planned creation by a Being capable of such planning and creating is the true explanation, care must be taken to identify and weigh up any arguments which might credibly support or contradict either of those two alternatives.

And I say this against the background that books that go to the trouble of arguing on behalf of *both* alternatives systematically and convincingly are extremely rare.

Taking everything relevant into account, the topic of this book has seemed to me to be an exceptionally interesting one. It is surely arguable, too, that it would be difficult to think of a topic that is more important.

* * * * *

We now have enough background to be ready to look at each of the two alternative theories of the origin of our universe in turn, the only remaining decision being: which of them shall we look at first?

I *could* just toss a coin, but there seems to be sufficient reason to give the theory of evolution pride of place, its being, worldwide, much the most widely held of the two theories.

PART II

On behalf of the theory of evolution.

Chapter 1
The theory of evolution introduced.

Let us start with an important question. How can we ensure that the case for the theory of evolution is satisfactorily presented in these pages? How can we ensure that it is presented as well and compellingly as it reasonably can be?

I doubt if we could do better than to invoke the assistance of one or more of the world's leading experts on the theory who have gone into print on it. Arguably the most important of the countless treatments of the theory of evolution that have been published is a massive book of 979 pages, *Evolution: The First Four Billion Years*, edited by Michael Ruse and Joseph Travis, published in 2009 by The Belknap Press, a division of the Harvard University Press, and consisting of sixteen chapters, each by a different author (in some cases two joint-authors). Worth quoting as an introduction to this PART II is the opening paragraph of the Foreword by Edmund O. Wilson:

> Two centuries after its author's birth and 150 years after its publication, Charles Darwin's *On the Origin of Species* can fairly be ranked as the most important book ever written... It is the masterpiece that first addressed the living world and (with *The Descent of Man* following) humanity's place within it, without reference to any religion or ideology, and upon massive scientific evidence that was provided across successive decades. Its arguments have grown continuously in esteem as the best foundation for human self-understanding and the philosophical guide for human action.

Also worth quoting is the opening paragraph of the Introduction by Ruse and Travis, the two editors of the book:

> The discovery of evolution is one of the greatest intellectual achievements of Western thought, ranking with calculus and

general and specific relativity among scientific discoveries that changed indelibly how we see our world. From seeing nature as fixed in form to seeing it as forever changing, we have been transformed utterly by discovering and understanding evolution.

Using that book by summarising its contents in order to present the facts and arguments favouring evolution-theory is an appealing idea on the surface, but unfortunately an impractical one. There is just too much material for it to be practical to condense it into this much shorter book that you are now reading, other than at the cost of doing so inadequately.

Fortunately, there is a book of many fewer pages than that of Ruse and Travis and both sufficiently complete and in every other way suitable for our purpose. *Why Evolution Is True* by Professor Jerry A. Coyne, a best-selling book published by the Oxford University Press also in 2009, covers all the relevant ground as thoroughly as could reasonably be wanted and has as its author someone of the highest relevant credentials. Coyne, who was briefly referred to in the previous chapter, is a professor at the Department of Ecology and Evolution at Chicago University, and the publisher's summary of the book tells us that he:

> is a specialist in evolutionary genetics, with his research focusing on the origin of new species, using the fruit-fly as his model organism;
>
> has taught evolutionary biology for more than twenty-five years;
>
> and, as important as anything for our purposes, has frequently contributed to public debate concerning evolution and creationism.

Clearly what should work very satisfactorily will be to make such use of Professor Coyne's book as will be helpful, and then look around in other authors addressing the subject, including of course those featuring in the just-mentioned and quoted-from

Evolution: The First Four Billion Years, in case there is anything left uncovered by him that can usefully be added.

Ideally, I would quote Professor Coyne exactly and extensively. I did not obtain permission when I asked for it, however, and therefore in what follows I am doing the next best thing, which is to represent what he says on those aspects of the subject as closely as I can without infringing his copyright. Please be assured that I have nowhere tried to weaken the case advanced by him. Far from it: if anything, it would have been my preference to *strengthen* it. Those interested in having confirmation that I am presenting his case to the best of my ability can of course check this by reading his book themselves. (When I last looked, it was obtainable very inexpensively at websites offering second-hand books for sale.)

Of immediate interest for evolution's case is that, in his Introduction of *Why Evolution Is True*, on page xix, Coyne makes highly critical observations about the general public in the Western world, and especially in the United States of America. He maintains that although he is able to assure his readers that the evidence for the truth of evolution is incontrovertible, opinion polls consistently show that Americans are suspicious of it.

For instance, when adults in no fewer than thirty-two countries were asked to give their views on the assertion that human beings "developed from earlier species of animals", a statement that he maintains is "flatly true, as we shall see", not more than 40% of the Americans to whom the question was put judged it to be true, even though the acceptance of evolution in many other countries was very much higher, 80% of them in some cases.

Furthermore, Americans pay even less respect to evolution when they make decisions on whether or not it should be a school subject. In a poll on the subject that was taken not long ago, about two thirds of them reckoned that, if evolution was to be taught in the classroom, creationism should also be taught.

Coyne said that, as an educator, he found it disheartening that a religion-based theory that had been discredited should be given the same status as a theory that was obviously true.

He rates the study and appreciation of evolution as important even just for the purpose of helping us to appreciate the world in which we live and our place in it. Truth on this subject is *surely* more mentally satisfying than the myth, formerly believed in by almost everyone, that we were suddenly brought into existence from nothing. In Darwin's words, quoted by him admiringly:

> **When I view all beings, not as special creations, but as the lineal descendants of some few beings which lived long before the first bed of the Cambrian system was deposited, they seem to me to become ennobled.**

Chapter 2

The theory of evolution as such.

Professor Coyne's first chapter in his book is titled "What is evolution?"

He opens by *appearing* to take the anti-evolutionist side. After pointing out that plants and animals seem to be perfectly designed for the lives they lead, he then says that what would appear to follow from this is that they were the result of the work of a master mechanic.

Where then does Darwinism fit in? He thinks it important to address this question because Darwin's theory, which in his opinion is profoundly beautiful when taken as a whole, is, he says, very often misunderstood.

What the theory of evolution holds is that life on earth began as just one tiny primitive species that, more than three and a half billion years ago, suddenly and by chance came into existence as something with life in it, and then, over a long period of time, branched out into other species which have continually increased in number.

If one ponders over that statement, Coyne continues, it can be seen that evolution-theory consists of the following components:

evolution itself;

gradualism;

speciation, which is the division of one species into further species by the evolutionary process;

common ancestry;

natural selection;

and various mechanisms of evolutionary change that do *not* depend on selection.

Having listed those components of evolution-theory, Coyne then looks at each one in turn.

The first of them is the actual idea of *evolution*, evolution meaning that over time a species changes its actual nature, *evolving* into different species of plant, insect or animal.

Gradualism, the second component, is what one would expect the word "gradualism" to mean. It addresses the fact that it must obviously take a very large number of generations of anything living for a really significant change – such as reptiles evolving into birds – to take place.

Under this heading, Coyne points out that all the countless living species of plant, insect and animal life have a number of things in common. An example is the *necessary* existence in all of them of the living biochemical "pathways" that enable them all to generate *energy*. This and other examples, he says, make it clear that *every* species of living things, including ours, ultimately goes back to a *single* common ancestor that itself had those features and passed them on to its descendants.

Thirdly, the reality that ten million species inhabit our world today, all of them arising from one ancestral form, requires, for its explanation, "the idea of *splitting*, or – a more accurate wording – *speciation*". This follows logically from the fact that, but for this development in evolution, there would today be only one single living species, a highly evolved descendant of the first species.

The primary result of speciation, thus defined, is that the original single ancestral species, which is often referred to as the "the missing link", eventually split into two separate descendant species, with this process then being repeated again and again. Interestingly, Coyne adds, more than ninety-nine percent of the species descended from this common ancestor have by now become extinct, leaving no descendants of their own.

Once we recognise the reality that everything living has a common ancestor, he continues, we are well placed to make important predictions about what is to be found in the fossil record, such as, for instance, common ancestors of birds and reptiles. Convincingly, predictable remains of such ancestors have indeed been found, providing us with some of most important evidence supporting the theory of evolution.

Fifthly, *natural selection*, which Coyne notes is what Darwin considered to be his greatest intellectual contribution on the subject of evolution. If inherited differences between one member of a species and another member of the same species result in one member being better able to survive and reproduce than another member, then the descendants of the better-surviving members of that species will continue to exist in greater numbers than the descendants of its other members. The result will be that this stronger species will fit in better with its environment, as its weaker members are gradually weeded out. For instance, hairiness will enable those who have it to survive better in cold climates than those that do not have it.

Sixthly and last is the fact that evolutionary change can, in addition, be caused by other evolutionary processes as well as by natural selection. Just to take one example, different families in the same species can have different numbers of immediate descendants, and these variations can cause changes *within* a species that have no connection with natural selection. That would explain, for instance, why sabre-toothed tigers went into extinction at a certain point in the world's history even though they had survived without difficulty for millions of years up until that time.

*   *   *   *   *

Coyne follows that summary of what he presents as the six components of evolution-theory by immediately then addressing the objection, "but evolution *is only* a theory, isn't it?"

The key word there, Coyne at once adds, however, is "only"; and the suitability of that word in the context of science depends entirely on the supposition that evolution is "a mere speculation and very likely wrong" – that is to say, equivalent to a guess.

What he goes on to say about that supposition is emphatic and unequivocal.

In the first place, what is involved there is actually a misunderstanding of the word "theory" in the present context. That is to say, in the context of science a theory is *very* much more than a mere speculation about how things are. Rather, it is a conscientiously-examined collection of propositions put together for the purpose of explaining facts about the real world.

In the second place, to qualify as a scientific theory, a theory must be capable of being *tested*, and in addition it must be possible to make use of it to make *predictions of future events* that in due course do indeed come to pass.

Interestingly, since a theory only becomes accepted as a true theory after what can be predicted from it has been repeatedly tested over a significant period of time, there is no exact moment when a scientific *theory* turns into a scientific *fact*. Rather, there comes a time when there is clearly enough evidence backing such a theory up for it to be recognisable as true by everyone capable of using his or her common sense.

Coyne then gives his readers a "rounding off" of this piece of, in the context, important scientific philosophy. In what he has just said he is very much *not* implying that true theories, recognised as such, are never shown subsequently to be in fact untrue. That can happen and has happened from time to time, and there is no signal to indicate to scientists that something that they have painstakingly arrived at as a reality of nature will never be overturned. Therefore, even though he maintains that Darwinism is now supported by countless pieces of accumulated evidence, it still remains a reality, he concedes, that something new could spring up to show that Darwinism is a false theory after all. That this will ever happen is improbable, to say the least, but, in Coyne's words, "scientists, unlike zealots, cannot afford to become arrogant about what they accept as true."

* * * * *

In the world in which *The Origin* was first published, almost everyone, scientist and layman alike, was a creationist, basing his or her view on how life came to be as it is in the early chapters of the Bible's Book of Genesis. What Darwin did in *The Origin* was offer evidence that *not only* favoured evolution *but also* was in opposition to creationism. Not that, at the time that he wrote *The Origin*, the evidence he was able to produce was decisive. The theory was therefore no more than a theory, however well backed up. Since 1859, however, more and more evidence has accumulated, and the theory has moved into the arena where, Coyne says, the word "facthood" can be validly applied to it. That is to say, just as the *theory* of gravity is a theory that is also a *fact*, so too is the *theory* of evolution in reality also a *fact*.

Creationism – the theory that life was created more or less as it is today, after having remained substantially unchanged ever since it came into existence – is of course still a popular theory. How can evolution-theory be tested against it?

There are, Coyne says, two kinds of evidence that are suitable for this purpose.

One of these is that of making use of the theory to predict what we can expect to find in both present-day-living species and extinct species of which there are fossil remains.

For instance:

– The fossil record ought to give evidence that there has been evolutionary change over the millions of years of history. We ought to be able to find some evidence of evolutionary change in the fossil record.

– The fossil record ought also to include instances of speciation, with a line of descent being divided into two lines of descent, which in turn are then subdivided into two or more further lines of descent.

– "Missing links" – more aptly called "transitional forms" – should be found occurring in layers of rock that date to the time when the groups are supposed to have diverged.

– Since evolution, very differently from creation, is very much a random development, there should in consequence be instances of imperfect adaptation.

– Finally, Darwinism is likely to be supported by what Coyne calls – it is a new coinage by him – retrodictions. These are the exact opposite of predictions. He defines them as effects in the natural world which, although evolution-theory may not make them predictable, nevertheless do not make sense *except* in the light of evolutionary theory.

Coyne develops further his thoughts on these "retrodictions" – these facts that support evolution as opposed to special creation because they can *only* be explained by evolution having taken place – and suggests that they include the following:

– patterns of how species *are distributed* on the Earth's surface;

– the existence of *vestigial features* that are of no apparent use, such as the wings of the ostrich and the kiwi and the eyes of the Mediterranean blind mole rat;

– even the human appendix, any benefits of which he reckons to be surely outweighed by the severe problems that can come with it, including the possibility of death from a ruptured appendix.

"Evolutionary theory then," says Coyne as he concludes the introductory chapter in which he analyses the concept of evolution, "makes predictions that are bold and clear." He then proceeds, chapter by chapter in his book, to look at the different categories of evidence that support it.

Chapter 3

The fossil evidence.

The first of the categories listed in the second chapter of Professor Coyne's book is *fossils*.

A fossil, Coyne says, comes into existence if and when the remains of a plant or animal become covered in water and then by sediment, with these taking place soon enough after the plant or animal has died to prevent deterioration of its remains. What has happened is that these remains have become compressed into rock by pressure of sediments piling up on top of them.

Often fossils are to be found in *layers* of rock, one layer on top of the next one. In such cases, it is reasonable to expect that fossils in lower layers are older than fossils in higher layers. It is, however, often far from easy to establish the time-order of layers because not all of them are present in any one place. It can indeed sometimes be necessary to connect layers that are in completely different parts of the world, and to "arrange them" according to the approximate date-order of when they came into existence.

Notwithstanding such difficulties, much has been achieved in this particular area of science, partly thanks to the ability to measure ages for some rocks with radio-carbon dating from 1945 onwards.

How reliable are such dating methods? This is a question of particular importance because, as Coyne says, people who do not accept evolution-theory do not hesitate to attack the reliability of this dating method by saying that supposed rates of radioactive decay cannot be relied upon. These rates might have changed, such people claim, whether simply with the passage of time or because of physical stresses that particular rocks have undergone.

This objection to carbon-dating, however, can easily be seen to be specious, he maintains. When dates that are given by radiometric dating can be decisively checked against historically recorded dates, as in the case of the well-known carbon-14 dating method, they are invariably in agreement.

Fossils are valuable for dating purposes, according to Coyne, not least because in the fossil record there are several kinds of evidence relevant to evolution.

One such piece of evidence is that fossils of early life are the *simplest* kind of species, while the fossils of more recent life are much more similar to those that still exist.

Another relevant piece of evidence for evolution-theory is that of fossilised organisms *splitting from common ancestors*, where the ancestors *themselves* can *also* be seen in the *fossil* record.

At this point, Coyne looks more closely at the relevant *history* shown by fossils. In chronological order:

– The first, and simplest, organisms, simple bacteria, are to be found in sediments that have been dated around three and a half billion years ago, which is only about a billion years after the formation of the Earth in which we live. For the next two billion years after they first came into existence, these cells were the only living things that occupied the earth.

– Then, as recently – if "recently" can be used in such a context – as about six hundred million years ago, a large quantity of relatively simple but multi-cell organisms arose, including worms, jellyfish and sponges.

– About four hundred million years ago, very simple four-legged animals appeared, and some of these are still with us.

– Around two hundred and fifty million years ago, the first mammals made their appearance.

– These were followed a hundred million years later by the first birds.

– Finally, only about seven million years ago, human beings, very much the newcomers in evolution-history, branched off from what Coyne calls primates, that term referring to the highest order of animals, such as apes, monkeys and lemurs.

Coyne then offers an interesting analogy. In his words:

If the entire course of evolution were compressed into a single year, the earliest bacteria would appear at the end of March, but we would not see the first human ancestors until 6 o'clock in the morning on 31st December. The golden age of Greece, about 500 B.C., would occur just thirty seconds before midnight.

In summary, now, of what Coyne has been telling us:

– The appearance of species through time, as seen in fossils, is far from random.

– Specifically, *simple* organisms evolved into *complex* organisms.

– Predicted ancestors evolved before descendants.

– The most recent fossils are those most similar to living species.

– We have transitional fossils connecting many major groups.

Neither the theory of special creation, he says, nor any alternative theory to that of evolution, can explain these very definite *patterns*.

Chapter 4

Speciation.

Speciation is Professor Coyne's next area of proof, falling under the heading of predictability.

What is needed in order to make a satisfactory examination of this aspect of evolution-theory is a good *succession*, historically speaking. Preferably, this succession will have taken place at a speed that makes change in any species relatively easy to see, and, preferably too, there will be no missing layers. For this purpose, very small organisms in the sea, such as plankton, are ideal.

According to Coyne, there are literally billions of such organisms, and, since many of them conveniently fall to the seafloor immediately after their death, their corpses have been piling up in a continuous sequence of layers.

There is no difficulty in examining these layers in the order in which they were formed. All that is necessary is to thrust a long tube down into the seafloor, pull up with the tube a column of what it has worked its way through, and apply dating methods to each part of that column. In instances where it is possible to trace a single fossilised species down through the column, its process of evolving can often be clearly identified.

The so-called "missing links" are *transitional* forms of organisms, and therefore of great value for the purpose of examining the process of evolution. These transitional forms are the fossils that span the gap between two very different kinds of living organisms that were the result of evolution.

They are of the utmost importance in terms of evidence, because, according to evolutionary theory, for every two species, however different, that have ever existed there was once a *single* species that was the ancestor of both – the so-called "missing link".

Coyne grants that the chance of finding any such single ancestral species in the fossil record is in fact, as he puts it, "almost zero". It is not, however, necessary for scientists to give up in relation to this aspect of evolution-evidence, because what *can* be done is to find *other* species in the fossil record that, being close cousins to the actual "missing link", are fully adequate for the purpose. And that has indeed been achieved.

He calls such species "transitional creatures". The dating of them, and to some extent their physical appearance as well, can be predicted from evolution-theory, as has been successfully done. Furthermore, some of the more recent and dramatic predictions that have been fulfilled relate directly to our own group, the vertebrates. Specifically – in relation to this piece of history of living things – up until about three hundred and ninety million years ago the only vertebrates were fish. Then, some thirty million years subsequent to that starting-point, and therefore dating back to as recently as three hundred and sixty million years ago, what are clearly *tetrapods* – four-footed vertebrates that walked on land – have been found. These early tetrapods were like modern amphibians in many ways, in that they had flat heads and bodies, a distinct back and well-developed legs; but they also showed strong links with earlier fishes.

How, though, Coyne asks rhetorically, did early fish evolve to survive on land?

That is where *prediction* has played an important role, he says in answer to his own question. A Chicago University colleague of his, Neal Shubin, used the logic applicable to evolution to predict that, if transitional forms existed, the fossils would be found in strata about three hundred and seventy-five million years old.

So it emerged. Shubin and his colleagues, after many years of searching in the Arctic Ocean north of Canada, came across a group of fossil skeletons stacked one on top of the other, which they could see was the transitional form that Shubin had been looking for. Shubin named it *Tiktaalik roseae* – "large freshwater fish" in the language of the people local to where he found it – since it had many features which show it to be a direct link between an earlier species of fish that had been identified and the later amphibians, animals that can live both on land and in water. On the one hand, scales and fins of *Tiktaalik roseae* showed it to be clearly a fish that lived its life in water. On the other hand, however, it had amphibian-like features, with its head flattened like the head of a salamander and its eyes and nostrils at the *top* of its skull rather than on the *side* of its skull, suggesting that it both lived in shallow water and was able to peer, and even perhaps breathe, above the surface of the water.

Coyne then moves on to another obvious problem, the problem of how it could have been that *birds* evolved. For instance, what possible use could only *half* of a wing be?

The answer, he responds, is not as difficult to arrive at as might first be expected. What must be put together is a series of intermediate stages of evolution of flight, each one useful to its possessor. Gliding is an obvious first step, and there are plenty of instances of that. One is that of flying squirrels doing this successfully with flaps of skin that stretch along their sides. Another, even more remarkable, is the "flying lemur", or colugo, of Southeast Asia, which, for gliding, has evolved a piece of skin on either side of it that stretches from its head to its tail.

If evolution be true, Coyne continues, we should expect to see transitions from reptiles to birds between seventy and two hundred million years ago. And transitions of that kind are exactly what *can* be seen.

The first such link between birds and reptiles was actually known to Darwin, and it is perhaps the most famous of all transitional forms, the crow-sized *Archaeopteryx lithographa*, discovered in a lime-stone quarry in Germany in 1860. *Archaeopteryx*, Coyne says, has both (a) some *reptilian* features, including a jaw with teeth, *and* (b) two *bird-like* features – large feathers and the kind of "big toe" that birds use for perching.

Other kinds of reptile-birds proved difficult to find for a long time, but in the late-1990s a series of extraordinary discoveries in China fully backed up the evidence represented by the *Archaeopteryx*.

There is still much that is not known in this area of science, Coyne accepts, but what can be done, even so, is to make educated guesses as to how modern birds evolved by means of natural selection. It is probable, for instance, that early dinosaurs evolved their longer front legs in order to be better able to "grab" their prey and then to deal with it. This "grabbing" would in turn cause muscles to evolve that would make their front legs stretch and also move inwards at the same time, which are the exact movements needed for the downward stroke of a bird's flight. That process would be followed by the birds acquiring the "feathery covering" that they have, which was probably needed for insulation.

As to how *flight* could have evolved from that point onwards: Coyne asserts that, while we can make educated guesses about the details, "the *existence* of transitional fossils – and the evolution of birds from reptiles – is *fact*" (my emphasis in both cases – N.M.G.). It is *fact* that *all* fossils of the just-mentioned *Archaeopteryx* and its descendants have features of both birds and reptiles, and it is *fact* that these occur at exactly the right time in the fossil record. Examples of what the fossils show are the alterations of *early* features, such as front legs with fingers, into *later* features, such as wings on which fingers are absent; and this is just what evolution-theory leads us to expect.

All such reptiles that have been found have feathers that make it possible for them to fly from tree to tree or from trees to the ground, and also help them to escape from stronger rivals, and to find food more easily, and of course to protect them if they fall.

At this point, Professor Coyne thinks it as well to say to his readers that he has included only a small selection of the examples of fossil evidence supporting evolution that he *could* have included, and he then gives further examples in summary, ending with what he considers to be the best example of evolutionary prediction being fulfilled: the human fossil record.

He follows that with a summary of the three most important facts shown by the fossil record.

First in this summary of what evolutionists claim: *the fossil record* makes it very clear evolution has taken place. Various objects that have been found in rocks give full confirmation of what the theory predicts, such as gradual change inside species before they become new species, descendants splitting up into different species, and transitional forms between organisms of very different kinds. This evidence is inescapable, incapable of being validly dismissed. "Evolution happened," he says in conclusion here; and he adds that, in many cases, we can see *how* it happened.

Second in his summary: such *transitional forms* as have been found occur in the fossil record *exactly* where they can be *expected* to occur, with, for instance, the earliest birds appearing *after* dinosaurs and *before* the birds of today.

Third: it is clearly indicated by the fossil record that evolutionary change, no matter how great, almost always involves the *"remodelling"* of existing features of living beings into new ones. Examples are the legs of land animals, which are variations on those of their fish-ancestors, and the very small middle-ear bones of mammals, which are variations of the jawbones of the reptiles they are descended from.

Such things, he says, can *only* be reasonably explained by evolution. A heavenly Creator would have no reason to create new species by means of the remodelling of certain features of existing species. Natural selection, by contrast, can only proceed by making changes to what already exists. Specifically, evolution-theory predicts that every new species will be a modified version of an older one, and that is exactly what the fossil record shows.

Chapter 5
Vestiges and embryology.

Professor Coyne follows the foregoing with a chapter titled "Remnants: vestiges, embryos and bad design".

He starts with vestigial evidence, and his first example under this heading is the ostrich. The *wings* of an ostrich, he says, are a *vestigial trait*: a vestigial trait being a feature of a species that has either lost its usefulness completely, or, as in the case of the ostrich, been co-opted for new uses.

We can, he continues, know that the ostrich is descended from flying ancestors, as are all birds that do not fly, both from the fossil evidence and from the DNA patterns that are always to be found in birds that do not fly. The wings have survived, but, although they can no longer be used for flying, they are by no means useless. What they do now is to help the birds that have them in such things as keeping their balance, mating, and even threatening their enemies. All flightless birds have wings, which either are simply remnants or, as in the case of the ostrich, are used for different purposes from their original purposes.

This, he says, answers the arguments of those opponents of evolution who say that so-called vestigial features are either obviously useful or useful for purposes that have not yet been discovered. Evolutionists do not claim that a vestigial feature has become useless, but claim, rather, that *it no longer performs the function for which it evolved*. Yes, the wings of an ostrich are useful, but it does not follow from this that they provide information relating to evolution. On the contrary, it is to be expected that features useful in the past will evolve into being useful in new ways.

Professor Coyne then turns his attention to us human beings.

Our bodies, he maintains, really teem with the remains of our ancestry from monkeys and other primates. We have for instance, the remains of a tail, a triangular end of our spine: that is to say, the coccyx, which consists of a number of vertebrae jammed together and hanging immediately below our pelvis. Another example is the tiny muscle attached to the base of each body hair which, when it contracts, makes that hair stand up, giving a "goose-pimple".

* * * * *

In the same chapter, Coyne then moves on to what he calls "bad design". In his view it is obvious that, if organisms which include bones, muscles, nerves and so on had been built up from scratch by an intelligent and capable designer, they would not have imperfections; and imperfections are therefore a clear sign of evolution, indeed just what evolution-theory would lead us to expect.

Under this heading, Coyne offers as his opening example the flounder, an overall term embracing the many different kinds of flat fish. The popularity, as eatable fish, of many of these various kinds of flounder, for instance the Dover sole, comes partly from their flatness, which makes it easy to de-bone them.

Flatfish, he then tells his readers, are born as normal-looking fishes that swim vertically, moving through the water with their heads pointing upwards and their tails pointing downwards, as though they are standing upright, so to speak; and they have an eye placed on each side of a pancake-shaped body.

A month after their birth a strange thing happens. One of the two eyes begins to move upwards, and then migrates over the skull and joins the other eye, so that there is then a pair of eyes on only one side of the body. Remarkably, the skull at the same time changes its shape in order to help this movement to take place, and there are changes too in both the fins and the colour. The flatfish then tips over onto the side that has just become eyeless and as a result *both* eyes are now on top; and, living on the bottom of the sea, it becomes flat and camouflaged and able and willing to prey on other fish.

Coyne's comment on that is that someone wanting to design a flatfish would produce a fish such as, for instance, the skate, which is flat from birth and lies on its belly. What such a person would *not* do, obviously, would be to produce a fish that needed to achieve flatness by lying *on its side*, requiring it somehow to move its eyes to different places and to deform its skull in the process. Flatfish are indeed poorly designed, he says, but this must come from their evolutionary heritage, rather than from an intelligent Designer, who obviously would not "build into" the fish such an illogical way of arriving at adulthood.

That gives rise to a problem which Coyne immediately addresses. Is poor design, he asks rhetorically, in fact an adequate argument for evolution? Coyne also reckons that embryology, the study of how fish, birds and animals develop in their different ways from the moment of their conception, offers important support for the theory of evolution.

Such a question misses the point, is his response. Certainly a designer may have motives that are unfathomable, but the *particular* bad designs that we see all around us make sense *only if they evolved from features of earlier ancestors*. Why would our hypothetical designer, when creating various species, sometimes do so in such a way as to deceive biologists by making organisms look as though they evolved?

Coyne also reckons that embryology, the study of how fish, birds and animals develop in their different ways from the moment of their conception, offers important support for the theory of evolution.

According to him, when biologists started studying embryology, along with comparative anatomy, their work turned up peculiarities that, at the time, did not make sense.

For example, all vertebrates begin their development in the same way, looking rather like embryonic fish. Then, from these starting-points, species that are *very* different in their forms and their appearances begin to emerge. Rather strangely: during this process, some of the organs in these vertebrates disappear while others undergo significant changes, and all this culminates in the dramatic differences that exist between fish, reptiles, birds and mammals, and also between vertebrates within those four groups.

This of course, Coyne says, raises many questions. For instance, why do the various kinds of vertebrates, ending up so different from each other in so many ways, begin their development with all of them looking like the embryos of fish? A second obvious question: why are the heads of mammals formed from the same "beginnings" as those which come from the gills of fish? And anyway, why does not natural selection eliminate the "fish embryo" stage of human development? After all, a fish-like circulatory system is hardly an efficient start in the direction of what will eventually be a human embryo.

It was, Coyne says, Darwin who pointed the way to recognising how these, to say the least, dramatic developments of various kinds took place. The probable answer, and a satisfactory one, is that descendants of an original ancestor inherited, and developed from, all the genes contained in the original ancestor, and, for a considerable period of time, they added, to quote Coyne, "new stuff" without getting rid of the "old stuff". The fewer changes there were from the earlier structures, the less likely that there would be deleterious "side effects".

This can only be a hypothesis of course, Coyne acknowledges; but the fact of embryology – with all vertebrates beginning their development looking like embryonic fish – can *only* make sense in the light of evolution having taken place and with all of us "cousins" being descended from a fish-like ancestor with a fish-like embryo.

When Darwin wrote *The Origin*, Coyne says, "he considered embryology to be his strongest evidence for evolution. It is sad," Coyne then adds when concluding this particular subject, "that, while embryology provides such a goldmine of evidence for evolution, textbooks on embryology often fail to point this out."

Chapter 6
"Biogeography".

Coyne's next chapter, titled "The Geography of Life", is about the distributions of plants and animals across the world. For this he coins a new word, *biogeography*, defined by him as the study of the distribution of species on Earth. For instance, why, he asks, were nearly all of Australia's native animals originally marsupials, while, in the r*est* of the world, it is the *other* kinds of animal that dominate?

Presumably, he says, because the biogeographic evidence for evolution is now so powerful, creationists act as though no such evidence exists. He has never come across any attempt to refute it in any a creationist book, article or lecture that he has come across.

It is obviously pertinent to ask why a creator would create different animals for different continents while nevertheless also giving them so much in common. No creationists, whether creationists who believe in the Book of Genesis or any other kind of creationists, have ever given any explanation that makes sense of why it is that different kinds of animals have great similarities in different parts of the world. They are simply reduced to invoking, in Coyne's words, "the inscrutable whims of the creator".

By contrast, evolution certainly *does* provide an explanation, which is the process called "convergent evolution". By this means, species that live in similar conditions will undergo similar selection-procedures, and therefore evolve in the same direction – they will *converge* – and thus they will become alike in many ways even though unrelated to each other.

An example of this convergence-in-action given by him is the white colours that are common to Arctic animals that are otherwise very different, as different from each other as the polar bear is from the snowy owl.

In evolving *convergently*, living things show the force of no fewer than three parts of evolutionary theory all operating together:

– One of these is *common ancestry*, providing the solution to such questions as why various kinds of marsupials in Australia share some of their features with each other, such as a double uterus, and other kinds of animal share other features, such as long-lasting placentas.

– *Speciation*, the second of the three parts, and defined back in chapter 2 of this PART II, is what causes living things with common ancestors to become so different from each other.

– *Natural selection* brings about the beneficial adaptation of each species to its environment.

Thus, in combination, *convergent evolution*: giving us, in its totality, a satisfactory explanation of this noteworthy pattern.

Returning to how the marsupials managed to find their way to Australia:

– The earliest marsupials for which fossils have been found, dating back some 80 million years, were located in North America.

– As these marsupials evolved, they moved southwards, reaching the most southern point of America some ten million years later.

– At that time, America and Australia were still joined together, as part of the then-continent now known as Gondwana (sometimes as Gondwanaland), with the tip of South America joined to present-day Antarctica and Antarctica joined to what is now Australia.

Accepting that this could be so invites a prediction: that it can be expected that fossil marsupials will be found on Antarctica.

Fired by this obvious possibility, scientists travelled to Antarctica in the search for marsupial fossils. They did so successfully. More than a dozen species of marsupials were found on Seymour Island, just off the Antarctic Peninsula; and they turned out to be of the right age, between thirty-five and forty million years old.

Chapter 7

Natural selection.

Professor Coyne then moves on to *natural selection*, which he calls "the Engine of Evolution".

Everywhere we look in nature, he says, we see animals that have the *appearance* of being actually *designed* to fit their environment. This applies to those parts of the environment that are physically relevant to them. Most obviously, it applies:

(a) to temperature and humidity;

(b) to other living beings that various species might come into contact with, such as predators that prey on them and species that they themselves prey on as predators.

It is far from surprising that naturalists of the past believed that such animals were the result of creation by God. In just one chapter of his *The Origin*, however, Darwin set aside the apparent need for a designer to explain this appearance of design everywhere. In place of a Creator, he offered *natural selection* – a completely unplanned, machine-like process that would bring about the same result.

Not, Coyne continues, that natural selection is without problems, when the theory is applied to biological processes. Is it really a sufficient explanation of, for instance, complicated adaptations?

Thanks to the efforts of dedicated biologists, there is no shortage of evidence that natural selection can and does operate in nature, and sometimes remarkably quickly. To show how it works, he offered an example: the adapting by wild mice of different coat-colours in accordance with different environments.

As of course we know, brown is the usual colour of mice, conforming to the colour of the soils in which they dig their burrows. On the coast of Florida, however, there are light-coloured mice, called beach mice, which are basically white, with a single light-brown stripe running down their backs. By an experiment – an unsentimental one, to say the least – conducted by Donald Kaufmann of Kansas State University, this was proved to be an adaptation for the purpose of camouflaging them from such predators as owls and herons.

Kaufmann put together a group of large cages, some of them with light brown soil in them and some of them with dark soil in them, and into each cage he put the same number of mice, divided equally between ones with dark coat-colours and light coat-colours. He then put into each cage a hungry owl, to see which mouse-colour was best for the survival of the caged mice. As might be expected, those with the fur that was most different in colour from the colour of the soil suffered the greatest reduction in number – proving the survival-value for animals of their being suitably adapted to their environment.

Since there are no other kinds of mice that are white, it can be assumed that these evolved from brown mice, with mice with a lighter coat surviving better than those more visible to predators than those with a darker coat – thus a good example of natural selection in action.

Chapter 8

Evolution's most impressive achievement?

Coyne's last-chapter-but-one, and the last chapter that we need to consider here, is called "What About Us?".

Darwin, Coyne informs us, came to the belief that we human beings originated in Africa, his reason being that Africa is where the animals most closely related to us, gorillas and chimpanzees, are to be found. No fossils supporting his belief have ever been found, however, and there is a very clear evolutionary gap between us and whatever common ancestor we may share with the great apes, an ancestor that must have been more apelike than humanlike.

In 1924, a first indication of a connection between us and gorillas and chimpanzees was located. Two boxes of rocks containing fragments excavated from a limestone quarry in Transvaal came into the possession of a young professor of anatomy in Transvaal's University of Witwatersrand, Professor Raymond Dart; and what Dart found in one of those boxes was the first specimen ever located of the remains of an animal – or was it an animal? – that he later named *"Australopithecus africanus"* ("southern ape-man"). Thus, for the first time, there was visible evidence that appeared to show that there was a definite link between us and gorillas and chimpanzees – what had been called up until then "the missing link".

Professor Dart spent three months carefully picking this piece of rock to pieces, and eventually he was able to see the full face of what had become fossilised – the face of a very small child, now known as "Taungs Child", that had identifiable milk teeth and erupting molars. Having located a mixture of both human traits and apelike traits, the professor was satisfied that what he had been given was an example of the very first of the ancestors of the human race.

Since then, other highly qualified scientists have made use of fossils and other information to show exactly where we are in history's processes of evolution.

And where exactly are we? Unquestionably, Coyne asserts, our original ancestors are apes, and those closest to us in the "ape-family" are chimpanzees, who, as just mentioned, split away from us several million years ago in Africa.

These, he assures his readers, "are indisputable facts", and we should find them deeply satisfying as such. Regrettably, however, he continues, that is not the attitude of everyone. There are creationists who accept that some species could have evolved from other species, but *all* creationists, without exception, stop short when it comes to human beings. Evolution, they say, cannot bridge the gap between us and the various kinds of ape, and from that it follows that an act of special creation must be involved.

In due course, fossil evidence was discovered that was sufficient to convince reasonably-minded sceptics that we human beings had indeed evolved. The result of Professor Dart's discovery of the Taungs Child was a search in Africa for human ancestors, and this eventually led to the discovery of "Lucy" by Donald Johanson in 1974, and also to many other discoveries, collectively giving us a fossil record showing how we evolved which, although far from complete, is considered to be definitely adequate for the purpose.

This is not to say that we have a continuous fossil record showing our ancestry. Crucially, however, when the hundreds of fragments that "Lucy" consisted of were put together, it could be seen that she – it turned out that the fossil was a female – was between twenty and thirty years old, and about three and a half feet tall, weighed some sixty pounds, and, as could be deduced from the manner in which the thigh-bone was connected to the pelvis at one of its ends and to the knee at its other end, *walked on two legs*.

Between her head and her pelvis, "Lucy" had a mixture of ape-like characteristics and human-like characteristics; and one could not reasonably ask for clearer evidence of transition between apes and human beings than "Lucy", since from the neck upwards she has a definite resemblance to an ape, from the waist downwards she is genuinely like a human being, and in the middle she is a mixture of the two species.

Coyne goes into considerable detail on various disagreements between evolutionists about how various stages of human evolution could have taken place, but insists that any remaining mysteries about the manner in which we evolved should not cause us to set aside the "indisputable fact" (his words) that evolve into what we now are is what we humans did. The certainty does not depend on fossils, because even without them we have enough evidence of evolution, as has been seen in the foregoing. The fossil record is, in his words, "really just the icing on the cake".

* * * * *

This is perhaps a convenient place to mention that, in addition to what Coyne has been telling us about *Australopithecus africanus"* and "Lucy", three other members of what is reckoned to be the *immediately* pre-human family tree have been – to use an appropriate term – unearthed in the past:

the so-called Neanderthal men, the first of which were found in the Neander Valley near Düsseldorf in Germany in the mid-19th Century;

the so-called Piltdown Man, claimed – deceitfully as it turned out – to have been found by Charles Dawson, an amateur archaeologist, in Piltdown, a hamlet in East Sussex in England, in 1912;

and the so-called Nebraska Man, found in the United States in 1922.

Chapter 9

Summing up with Professor Coyne.

The final chapter in Professor Coyne's very useful book is titled "Evolution Redux", and in it he rounds off his presentation of his case.

It is his constant experience, he says, that, convincing though the evidence supporting evolutionary biology very evidently is, many people, although unable to refute the evidence, fail to be convinced that evolution-theory is true. Such problems do not exist in other areas of science, so what can be the specific problem with evolution?

Certainly it is not lack of evidence. As he claims to have shown in his book, evolution is not only a mere scientific *theory*. In his words: "It is a scientific fact." All the evidence that he has put before his readers, such as the fossil record, "biogeography", embryology (embryology being the science of how an animal develops from the moment that it is conceived), the vestigial structures, and less-than-perfect design, show "without a scintilla of doubt" that living things have evolved. Natural selection has actually been seen in action, and there is every reason to suppose that complicated living beings can be the result.

Undoubtedly, he firmly states, the major features of Darwinism have been shown to be true. Organisms naturally and gradually evolved, and then turned into different species, with natural selection being the principal cause of the organisms adapting to their various environments. "No serious biologist doubts these propositions."

It could hardly be put more strongly than that.

Chapter 10
Is there anything to add?

Back in chapter 1 of this PART II, I said that we would start by making such use of Professor Coyne's book as could be seen to be helpful, and that we would then look around to see if any aspect of the subject left uncovered by him could usefully be added with the help of other authors on evolution.

I for one am always interested in evidence that could point in *either* direction, whether in favour or against evolution, and over the years I have accumulated a large number of books covering the subject. Very evidently, the most important of those favouring evolution is the 979-page *Evolution: The First Four Billion Years* by Messrs. Ruse and Travis, referred to in chapter 1. Between them, that book and Professor Coyne's *Why Evolution is True* give all the support for the theory of evolution that could reasonably be needed in order to make as good a case as anyone could hope to become acquainted with. In summary of what Coyne and other authors have set down before their readers, the basic case consists of: –

Fossils that have been found in sedimentary rocks and, in vast numbers, in fossil "graveyards" on land in various parts of the world, from the United States to China, and the supposition that ape-men eventually evolved into human beings…

Mutations, sometimes called speciation, giving rise to various different breeds of rabbits and to various different races of mankind.

Radio-active methods of dating the earth.

Carbon-14 dating of coal and oil, both of which take many millions of years to form.

Biogeography.

Vestiges in various creatures that are still part of the world around us, including of course ourselves.

* * * * *

Are there any arguments that proponents of the theory of evolution have offered or can offer in addition to the foregoing?

Quite recently, a new book of evident relevance has appeared in the bookshops, *Outgrowing God – A Beginner's Guide* by the best-selling atheism-and-evolution-promoting author Richard Dawkins, who was Oxford University's Professor for the Public Understanding of Science until his retirement at the age of nearly seventy in 2008.

I really do mean best-selling, incidentally. One of his books alone, *The God Delusion*, has generated sales of more than three million copies, an extraordinary figure for a work of non-fiction. And that, as I say, is just *one* of his books.

Such a book by such an author is obviously worth giving consideration to, and all the more so in this case because it is arguable that *Outgrowing God – A Beginner's Guide* is in some respects even more interesting than Professor Coyne's book, in that – a rarity among proponents of evolution-theory – Professor Dawkins does at one point set out to examine and answer the best arguments against evolution, in a chapter, chapter 7, called "Surely there must be a designer?".

At first sight, he gives the impression there of making a satisfactory job of presenting the anti-evolution case. He opens the chapter with a description of a cheetah chasing a gazelle in the African Savannah, and, after going into considerable detail on what is involved physically in this, for both of those two animals, makes the observation that "both cheetahs and gazelles seem superbly 'designed'." After elaborating on this in some detail, he rhetorically asks how the complexity in the construction of both of these fast-running creatures can have come about:

Must they have been designed by a mathematically-minded genius? The answer is an emphatic, if surprising, no – and we'll see why in the following chapter.

He then invites us to think of the eyes of both the cheetah and the gazelle, describing them along the same lines as the description of the eye that will be given in chapter 4 of PART III of this book, and making the point that in many respects the eye of any animal resembles a carefully constructed camera.

After that, he describes in wonderment the extraordinary workings of the tongues of chameleons, and also the ability of chameleons to change colour to match the background, and then offers the comment:

> Once again, this all looks as though it demands a designer, doesn't it? Once again, it really doesn't, as we shall see in the next chapter.

Another example that he gives is the human brain – even more amazing, he says, than any eye, containing as it does a hundred billion nerve cells that are "wired up to each other in such a way that you can think, hear, see, love, hate, plan a barbecue, imagine a giant green hippopotamus or dream of the future."

After giving several more examples of apparent design, he then changes tack and gives examples of what he refers to as "bad design", under which heading he includes the flounders described and discussed by Professor Coyne as noted earlier, in chapter 5 of PART II of this book. This species of fish, he points out, lies on either one of its sides, rather than on its belly, creating the problem for it that its eye facing downwards against the bottom of the sea is "pretty useless".

And what did the flounders do about this? "They grew a distorted, twisted skull, so that both eyes can look upwards instead of the view of one of them being restricted to the sea bottom. And I do mean twisted and distorted. No sensible designer would have produced an arrangement like that."

He dismisses the notion that God, if He existed, would have included in his creation such examples of bad design, and then maintains that, on the other hand, Darwinian evolution by natural selection "explains it – and everything else about life – perfectly well", as he will be showing.

In the next chapter, chapter 8, the first of the later chapters in his book that he has been constantly referring to in the foregoing, Dawkins first makes the point that it is obvious that animals and plants do not come about by random chance, and then puts the question: "So, what is the alternative?"

His immediate response to this question is: "Unfortunately, at this point, many people go straight down the wrong path. They think that the only alternative to random luck is a designer."

That indeed is what almost everybody thought, he continues, until Darwin came up with his revolutionary idea. What almost everybody used to think, however, "is wrong, wrong, wrong."

He then proceeds with examples in which he shows what he reckons to have *really* happened in the various instances.

For instance: "Suppose," he says, "a cheetah is born with claws just a tiny bit longer than in the previous generation." This completely random tiny change, he maintains, will probably either make the cheetah a little worse at surviving or a little better. If better, it will not only *itself* survive better. It will *also*, in at least some cases, pass on its slightly longer claws to the next generation. Somewhere in the make-up of this cheetah there is a gene that affects the length of its claws, and its descendants will inherit this gene, to good or bad effect; and, in the case where the result is that an animal or plant is more likely to survive as a result, natural selection will eventually make the lengthening of the claws a permanent feature of the cheetah.

Thus what Dawkins speculates to his readers, as support for his conviction.

He follows this example with several other examples that he believes illustrate the same point, and then, further developing that point, adds that, if it only takes a few centuries to turn a wolf into a whippet, "just think how much change could be achieved in a million centuries."

Next he returns to the eye. Although, obviously, the notion that something like the human eye could spring spontaneously into existence is a hopelessly unlikely one, what *can* happen without offending against reasonable probability, he says, is that "an excellent eye can come from a random change to a slightly less excellent eye, and that slightly less good eye can come from an even less good eye, and so on back to a poor eye."

And of course even an extremely poor eye is better than no eye at all; and the same is true of legs and hearts and tongues and countless other parts of bodies – all of which leads to the conclusion that "everything about living creatures, no matter how complicated, no matter how improbable – can now be understood."

The key point, as far as Dawkins is concerned, is that what at *first* sight appears to be *hopelessly* improbable becomes *not at all* improbable when it is recognised that it can arrive "gradually, stealthily, step by tiny step, where each step brings about only a really small change."

He then describes the role played by DNA,

> A genetic molecule whose importance is almost impossible to exaggerate

and which consists of, in effect,

> a chain, a necklace of jigsaw pieces that are chemical units called nucleotide bases... Ultimately, it was our DNA that determined how each of us developed from a single cell into a baby, and then grew into what we are now, the DNA being, in effect, a set of instructions for how to build a baby.

And the point is that *any* mutation – any change in DNA in any plant or animal – can have the effect of making it more or less likely to survive, and more or less likely to reproduce as well. And, "as the generations go by, over thousands and millions of years, the genes that survive in the population are the 'good' genes," enabling bodies to run faster or have longer lives, or whatever. "That, in a nutshell, is Darwinian natural selection, the very reason why all animals and plants are so good at what they do."

* * * * *

I believe us now to be adequately acquainted with the evidence, of many different kinds, that supports the theory of evolution.

PART III
Are there, however, any problems with the theory of evolution?

PART III

Are there, however, any problems with the theory of evolution?

Chapter 1
Introductory.

For complete clarity about what is under discussion, let us once again address this question, put to us by Professor Coyne in PART II: What exactly *is* evolution, as represented by evolution-theory?

Rather than risking even the smallest element of inaccuracy by using my own words, I offer the best summary that I have come across, given in a book written in the 1920's, *After Its Kind* by Byron Nelson:

> "Evolution", as the word is used in the widespread discussion of the present day, denotes a process which has taken place entirely naturally, without the miraculous intervention of any Divine Being, by which, from out of a single remote ancestor living in the waters of some distant sea, have come all the living things in the world today. It is a natural process which, if it ever took place, would enable all birds, fish, reptiles, mammals, apes and men to trace their ancestry back from all directions to a speck of protoplasm that somehow came into existence hundreds of millions of years ago.
>
> "Evolution" means a process by which man must trace his ancestry back to some ape form, then to some quadruped, thence to some reptile, thence to some amphibian, thence to some single-celled creature that lived in the slime of the sea. If "evolution" is a fact, then species have never been fixed and are not so now, but have been continually drifting over from one form into another since world history began. This is the commonly accepted meaning of the term.

As we learnt in PART II, it is not as a mere *theory* that the theory of evolution is commonly presented to the world, by scientists, academics, and the communications media; in schools and museums; in books ranging from light fiction to scholarly tomes. Rather, in spite of what the word "theory" would suggest, and even dictate, evolution-theory is presented as *fact*. Granted, there is considerable disagreement about (a) *how* evolution took place, (b) *how much time* it took, and (c) *the various forms* that it took. Occasionally it is admitted that all that we can see from our own experiences is *the very opposite* of evolution: that is to say, degeneration. Nonetheless, it is never doubted, either by atheists or, other than in certain areas in the United States, by the majority of those who hold religious beliefs, (a) that evolution happened, and (b) that evolutionary theory, with the help of compatible theories in such fields as astronomy and geology, provides the only possible explanation for how the universe, the world and the creatures in the world came into existence.

Let us now, even so, take a critical look at the most important arguments put forward by evolutionists. In doing so, I shall be listing and examining a number of absolute and undoubted laws of nature that appear to demonstrate *conclusively* that the theory of evolution is actually impossible from many points of view.

I shall also be examining with some care the most important arguments that those who believe in evolution use to justify their beliefs. "Mountains of evidence," there is for evolution, said Professor Coyne on page xiv of his book *Faith versus Fact* that he wrote subsequently to his *Why Evolution is True*. I shall try to make sure that no evidence of any significance is left unanswered, at least by implication.

Incidentally, please note that, other than in the case of the titles of books and periodicals, the italicisations in the quotations in following pages are always mine.

Chapter 2

Can one species change into another species?

The first question to be addressed is very evidently fundamental to the theory of Evolution. It is: can one species *ever* turn into another species?

As a preliminary to answering this, it is important is to take note of what the term "species" actually means.

In the words of Byron Nelson, whom I quoted in the last chapter, on page 8 of the same book:

> **The scientific definition of a species adopted and defended (in *Nature*, 15th July 1922) by Professor William Bateson, President of the British Association for the Advancement of Science 1914-1927, is as follows: "A species is a group of organisms with marked characteristics in common** *and freely interbreeding.*"
>
> **This definition allows for the variation which we know exists in natural species, and yet acknowledges the existence of the wall of partition** *between* **natural species known as sterility, the true test of natural species.**

In other words, varieties of living things *within* species – even hugely different varieties – can interbreed. With very rare exceptions, members of two *different* species are unable to. That indeed is exactly what is contained in the very definition of the word "species" just given by Professor Bateson.

Dear reader, this may come as a surprise to you, but please take careful note of the following, where I quote A. N. Field on page 13 in his book controversially titled *The Evolution Hoax Exposed* (Christian Book Club of America, Hawthorne, Ca., 1971):

There is, unfortunately for evolutionists, *not a shred of evidence of any living thing ever evolving into some different kind of living thing that is capable of breeding but is infertile with its parent stock.* All that breeding experiments have produced is mere varieties that are fertile with their parent stock, or else sterile hybrids incapable of breeding, such as the mule produced by a cross between horse and donkey. All living things go on obstinately producing young according to their own kind *and to no other kind.* Evolution has to show that living things can break through their natural breeding limits. And this is just what evolution has been quite unable to show.

What is worse still for evolutionists, at about the same time as Darwin was formulating his theory, a monk called Gregor Johann Mendel was working on the laws of heredity in nature and making a discovery, now known as Mendel's Law, which showed *the exact opposite* of Darwin's theory. He demonstrated conclusively that we are what we are because of what our parents were, and that offspring inherit characteristics from parents. That is to say, exposure to a new environment is not sufficient to give a member of any species new characteristics – characteristics that did not already exist in the species.

I now turn to a revealing passage in another book, *The Rise of the Evolution Fraud* by Malcolm Bowden (first published by the author in 1982 and republished in a revised and enlarged edition in 2008):

Claims [on behalf of evolution] are made from time to time of the production by experiment of new species of living things, but they rapidly drop out of sight. This evidence is vital to the evolution-theory, and if it were forthcoming we may be quite sure it would be proclaimed from the housetops for all the world to hear.

If this evidence is lacking, it is not for want of seeking it. For example, a whole literature, so extensive that a bibliography of it was recently published, has grown up about the breeding experiments with the pumice fly (fruit-fly), *Drosophila Melanogaster*. Mr. Douglas Dewar, a Fellow of the Zoological Society and one of the few British biologists rejecting evolution, on pages 20 and 21 of his *Challenge to Evolutionists*, relates how in 1910 Morgan and his collaborators hit upon the idea of experimenting with this quick-breeding fly.

This obliging little creature produces 25 generations a year at ordinary temperatures and more at higher temperatures. More than 800 generations of it have been bred with the object of transforming it into something that is *not* a Drosophila Melanogaster. It would take 20,000 years to get as many generations of human beings. Every device has been applied to this fly to make it vary its breeding. In 1927 it was discovered that by exposing it to X-rays the rate at which mutations, or marked variations, occurred could be increased by 15,000 per cent.

These breeding experiments are stated to have resulted in the production of some 400 varieties of this fly, some of them monstrosities, and some differing more from the parent form than the other wild species of Drosophila differ from one another. Nevertheless, all these varieties (unless they are too imperfect to breed at all) are stated to breed freely with the parent stock, whereas the *different* wild species of *Drosophila*, on the rare occasions when they can be induced to cross, yield either no offspring at all or sterile hybrids. Immutability of species, like a mysterious angel with flaming sword, stands barring the way to the evolutionists' Garden of Eden.

Summed up, the position is that there is no evidence of any interbreeding community of living things being able to change its breeding and become transformed into some different kind of thing, and infertile with the original stock. Evolution asserts that *all* species came into being in this way. And evolution is wholly unable to provide *any* vestige of proof of its assertion. Belief in evolution today must thus rest on 'general considerations', just as Darwin privately confessed was the case way back in 1863.

Scientists justify their belief in evolution by relying on what they think *must* have happened. That, however, is not science. Science is knowledge, not mere supposition; and to prove something *scientifically*, more, very obviously, is needed than assumptions.

Put in simplest terms, a dog mating with a dog is *only* going to produce a dog. The mating of a stallion and a mare of the horse species will *only* produce horses. Yes, as mentioned earlier, there are instances when two species can mate and produce a cross – the horse and the donkey producing a mule, or a lion and a tiger producing a so-called liger, for example. But their offspring are then infertile, or, very rarely, revert to either one or other of the original animals. Nature, "refuses" to accept developments that could lead to a new species, and in fact weeds them out.

The changing and varying of attributes *within* species – dogs with thicker coats or flatter noses, horses being bred for racing – is a reshuffling of existing genes, either to adapt to a specific environment or to cater to human fancy. This certainly does takes place and we see it all around us. On the other hand, however, a leap of one species into another, through new genetic information being introduced into the gene pool, has never been seen.

* * * *

Failing to produce a living creature (if I may use that term) that can breed within its species but is infertile with some of the other descendants from the same parents is not the only problem that breeders have failed to overcome.

By selectively breeding freaks, mating freak to freak, and culling those that do not conform to the breeders' policy, it is possible (always at a cost, if only in health or hardiness) to produce animals with organs that are more highly developed in certain respects than the norm – to produce, for example, an exceptionally fast horse suitable for racing, or a dog with a keener than average sense of smell, or a cow that produces a vast quantity of milk. It is possible also to breed animals *without* certain properties with which the species is usually endowed by nature, such as cows without horns.

One thing, however, is *not* possible. No breeder has *ever* produced an animal with a new limb or organ. However useful he might find it to breed a cat or a horse with a pair of horns or a third eye, even common sense alone is sufficient to show that he would try to do so in vain.

Certainly, substantial mutations do take place from time to time, normally because of some form of poisoning during the period of gestation. For instance, both animals and humans are occasionally born with an extra, but useless, limb. Such mutations, however, are *invariably harmful*, and *reduce* the likelihood of the survival of the species rather than improve it. They are, in fact, the result of genetic *injury*, and they would not make it possible for a human to be born with, for example, gills or wings, which are not part of its genetic blueprint.

"Mutation theory", as it is called, had in fact been offered as a serious explanation of how evolution must work at around the beginning of the twentieth century by the botanist, Hugo de Vries. While experimenting with the breeding of primroses in his garden, he noticed that occasionally there appeared new forms of primrose that he had never seen before. He labelled them "new species" and came to the conclusion that evolution, instead of coming about very gradually with each addition being so tiny as to be unnoticeable, arose suddenly and spontaneously by steps and jumps.

Mutation theory had to be discarded, however, when it was discovered that the specimens of de Vries's supposedly new species that he had identified *were already in existence*, and that his various "new species", so-called, were no more than varieties of forms of primroses of which he had previously been unaware. Some varieties of species remain dormant until brought forth by the right combination of genes coming together; and it was such varieties, *still* within the same species, that his experiments had produced.

* * * * *

Let us now look in a little more detail at some of the problems that evolution-theory unavoidably gives rise to.

We are doubtless all aware that Charles Darwin advanced the theory of:

"natural selection"

which arises from

"the survival of the fittest"

in

"the struggle for existence".

As will be seen when we come to look at that theory of Darwin's in more detail, however, it does not stand up to any serious examination. Selection can *only*, by the very laws of nature, *take away* from what *already exists*. It cannot *produce* what does *not* already exist.

In other words, the *arrival* of the fittest is needed before any *survival* of the fittest can take place, and natural selection can *only* deal with *what is already there*. Selection, by the very nature of things, is *not* about bringing *anything new* into existence.

We return once again to Byron Nelson's book *After Its Kind*, this time on page 94:

Nature selected the more fit to survive and the less fit to perish, as Darwin said. But... we have not yet had explained to us by Darwin how those parts of the body came to be. *This* is what we want to know. The views of a few keen-thinking men on this weakness in Darwin's theory may be noted:

Professor Lock of Cambridge University: "Selection, whether natural or artificial, can have no power in creating *anything new.*" (*Variation and Heredity*, p.40.)

Hugo de Vries: "Natural selection may explain the *survival* of the fittest, but it cannot explain the *arrival* of the fittest." (*Species and Varieties*, pp.825-6.)

Alexander Graham Bell, the inventor of the telephone and a researcher into evolutionary problems: "Natural selection does not and cannot produce new species and varieties. On the contrary its *sole* function is to *prevent* evolution." (*World's Work*, Dec. 1913, p.177.)

Professor Coulter of the University of Chicago: "The most fundamental objection to the theory of natural selection is that it cannot *originate* characters; it only selects among characters already existing." (*Fundamentals of Plant Breeding*, p.34.)

And what comes next is from a much more recent book, *Evolution: The Greatest Deception in Modern History* by Roger G. Gallop, Ph.D., published in 2014 (first edition 2011), page 29:

> Natural selection or adaptation – commonly known as "survival of the fittest", speciation or micro-evolution – is the genetic process within a "kind" of animal population which selects gene traits from an existing gene pool best suited for a specific environment.

In other words, "Nature" can naturally select dogs and other animals by "survival of the fittest" – this is nature's way of allowing animals to adapt to their environment. In a cold region, for example, dogs with thick hair will be more likely to survive, and their genes will be passed down. No *new* genes have been added, however, and there is therefore no evolution.

I now refer back to Professor Richard Dawkins' book *Outgrowing God – A Beginner's Guide*. In that book he *nowhere* answers, or even brings up for any sort of consideration, the *fundamental* objection that, at the very most, natural selection can only improve *what already exists*.

* * * * *

Domestic species introduce another interesting problem for evolution-theory believers. A little-known feature of all domestic species is that their step-by-step ancestry back to wild animals has never been traced. Indeed it is not possible to prove that they *have* such an ancestry. Here is *The Catholic Centre for Creation Research* on the subject, in its *Newsletter*, volume I, 8th March 1976, page 6:

> The special creation by God in the beginning of both domestic and wild animals is confirmed by anthropological and archaeological studies. The earliest civilisations had domestic animals. The derivation of domestic animals from wild ones during early Neolithic or back into Mesolithic times is evolutionary speculating and wishful thinking.

To supplement that, here is John Kasper, on page 36 of his strangely-titled but excellent book, *Gifts from Agassiz*:

> Our domesticated animals, with all their breeds and varieties, have never been traced back to anything but their own species. Nor have the artificial varieties, so far as we know, failed to revert to the wild state when left to themselves.

The reality is that all attempts to take a breed of wild animals and turn it into a breed of domestic animals have failed. Remarkably, this is so even in the very rare cases where domestic and wild animals have the capacity to interbreed, thereby suggesting more strongly than in other cases the possibility of a common ancestor. For instance, wolves have interbred with dogs and the resulting progeny and their descendants are immediately domesticable. *If*, however, the same wolves are kept in captivity and are no longer interbred with dogs, they do *not* become domesticated, no matter how many generations the captivity lasts for. Although, therefore, there may be, through interbreeding, a certain proportion of wolf blood in many or even all breeds of dog, neither is the wolf the direct ancestor of the dog, nor can any theoretical – but in fact only mythological – ancestor of the wolf be the common ancestor of both.

Chapter 3
Laws of nature.

What we are now going to look at are clear laws of nature and therefore of science – *laws*, not theories – that *very* obviously govern the natural world. Even just *one* of these laws of nature would be sufficient to eliminate any possibility that Darwin's theory could have any validity. Two such laws would be even more sufficient of course, if impossibility were to admit of degrees. And the reality? There are no fewer than *eight* laws of nature that rule out the theory, laws that are so *self-evidently* laws that *anyone*, of *any* degree of learning, can recognise that they are indeed laws.

There is of course an essential and fundamental difference between a law and a mere theory. A theory is a proposition which is worth considering unless and until conclusive arguments can be seen to oppose it. *A law of nature*, by contrast, is a proposition which admits no exceptions. *The moment* that an exception was found, it would cease to be a law.

Against that background...

Law 1: *The law of entropy,* also known as *the second law of thermodynamics.*

First, the context of this second law of thermodynamics. The first of the two laws of thermodynamics says that heat and energy cannot be created or destroyed. Heat and energy can only flow from place to place or change from one form into another form. Within this law, entropy is a *quantifiable* measure of how *evenly distributed* a given amount of heat or energy is.

The second law of thermodynamics states, amongst other things, that:

(a) as time elapses, *order can turn into chaos*, and inevitably does so, if any change takes place at all, and does so increasingly as more and more time passes;

(b) but *chaos can never turn into order*, no matter how much time is allowed for.

Yet the world that we live in is of course is full of "order", often very complex and detailed order. And the law of entropy dictates that this order, is in the extent of it, inescapable evidence of the most careful planning.

The second law of thermodynamics then goes on to say that, within a *closed system*, which is a system that does not exchange any energy with the surrounding environment, entropy always *increases*, which is another way of saying that the energy will always be more widely dissipated and therefore increasingly less capable of doing useful work. This basic principle applies throughout the physical world. Thus: –

Walls made of bricks tend to fall apart; loose bricks do not fall into place as walls.

A tidy room in use that is not kept tidy will become untidy, and an untidy room in use will never become tidy unless an orderly plan to tidy it is carried out.

Everywhere, order tends towards chaos; never chaos towards order.

And so on.

In other words, the material universe cannot *ever* have been becoming more creatively *organised-in-appearance* and *orderly* than before. On the contrary, it can *only* have been always running down. This is sufficient to make it impossible that the universe could be eternal, for instance; because entropy – and the lack of any energy whatever in a form capable of useful work – would long ago have been complete in an eternal universe. Indeed, at the rate of decay of heat, as can be easily measured today, entropy would have been complete within a few hundred thousand years at the outside, let alone by the end of the few-billion years so far postulated – indeed supposedly proved – by the scientific community. Yet of course it is in reality *far* from complete.

It is worth noting that the law of entropy operates in exactly the *opposite* direction to the commonly-accepted Big Bang theory and evolution-theory, whether we consider them individually or in combination. This is a fundamental and overwhelmingly obvious point, and cannot do otherwise than rule out evolution that is unplanned and random.

Law 2. This law is conveniently summed up by a Latin tag: "*Ex nihilo nihil fit*" (translation: "Out of nothing, nothing [is] can be made"). It simply says that, *out of absolutely nothing, ONLY absolutely nothing, can possibly come.* That is obvious enough once stated, and what we are therefore left with is the question: where did the universe come from?

For the universe to have made itself, it would have to have existed in order to do the making, which is self-contradictory.

The only other alternative possibility, if we leave aside that of an all-powerful Designer, is that the universe is "self-existingly" eternal. Just to mention one absurdity that must be skipped over in order to believe this: as mentioned under the heading of law number 1, all around us we see evidence of things running down, wearing out. Even the sun is losing energy. If this had been going on from the "beginning" of eternity, everything that could run down would -- long, long ago -- have reached the end of its ability to run down and to go on expending energy. The sun would have burnt out, down to nothing, aeons ago.

It was to avoid having to face up to the problems of this law number 1 that the "Big Bang" theory was produced, a theory now almost universally accepted as fact by scientists. Not only, however, is there no shred of non-imaginary evidence for it, but belief in this theory requires belief in the following:

(i) that, some 13.7 billion years ago, the entire mass of the universe was compressed into a single, almost impossibly-tiny point – to give it its technical name, a "primeval atom";

(ii) that this almost impossibly-tiny point was *extremely* hot and *extremely* dense;

(iii) that this minute "thing" was suddenly caused to expand by its extreme heat, and has been expanding ever since and losing heat in the process, resulting, particularly at our point in its history, in the temperatures of the various parts of the universe as they are now;

(iv) that it is, moreover, still expanding.

Seriously, this really is what most of today's scientists believe, as can be verified by looking up Big Bang theory in any standard source.

Where, however, can this remarkable "minute point" have sprung from? How can something minutely small as this "thing" supposedly was, produce even something the size of a small pebble, let alone what we see above us and all around us? How can something as haphazard as an explosion be responsible for the extraordinary *orderliness* that such scientists as astronomers, physicists, chemists and biologists recognise, and which indeed makes the study of their subjects possible?

How *can any* of this be considered possible, other than by today's "scientists"? – or dare I say "… *even* by today's scientists"?

To those searching for answers, I draw attention to this exact quotation – and I emphasise that it really is an *exact* quotation – by two highly regarded present-day authorities on science:

Because there is a law of gravity the universe can and will create itself from nothing.

83

That was said jointly by Doctor Leonard Mlodinow and the late, as of 2018, Professor Stephen Hawking on page 180 of their book *The Grand Design*, published in 2010.

What, arising from that statement, we are asked by these authors to believe on their say-so is that laws can exist without anything existing in nature to be laws *of*. Laws, moreover, which actually have *creative power*, infinitely more creative power than human beings of the highest intelligence and competence can have; power indeed that is sufficient to enable those who have it to create *a whole universe* out of *nothing*.

The reality is that laws of science can tell us how things in existence *act*, including becoming transformed into other things, as water can become steam. What, however, laws of science can *not* tell us is how, in whatever combination of laws, things such as even the most ordinary small pebble can be *created* – brought into existence – out of absolutely nothing.

Law 3: The ordinary law of cause and effect. This law says *that nothing can happen without a cause that is sufficient to bring about the "happening", and that nothing can exist without a sufficient reason for its existence.*

To take just one example of this law not being adequately faced up to by all the most respected experts: –

In the first place, there is everywhere in the universe evidence of design.

In the second place, design cannot take place without deliberate and intelligent planning and execution. Indeed the expression "intelligent design", sometimes used by scientists in this context, is incompetent language and muddled thinking. Specifically, it is tautologous, because the concept of intelligence is already included in any adequate definition of the word "design".

There will be more on this subject when we come to the next chapter.

Law 4: A variation of the law of cause and effect says that *a cause must not only exist to produce a given effect but must be adequate for that purpose, although not more than adequate.*

But the human brain, for instance, can do vastly more than is needed for mere survival. Building the Taj Mahal; composing music; playing and dancing to the waltz-melodies of Johann Strauss; writing the works of Shakespeare; programming computers; putting together these notes for your attention – *how* can any of these contribute to survival?

Law 5. This law can be conveniently summed up by the Latin tag, *"Nemo dat quod non habet"* – *"No one can give something that he has not got."* We cannot give *anything* unless we both (a) possess whatever it is that we wish to give and (b) have it under our control. This law applies at all levels. I can give you neither a gold coin if I have not got one, nor even a good idea if I have not got one.

Equally, and essentially relevant to our present discussion, two parents cannot pass on to a child anything that neither parent possesses in his or her genes. Indeed all selective breeding, from racehorses to plants of any kind, completely depends on this principle. And Mendel's famous law referred to back in chapter 2 of this PART III – note that it is referred to as a *law*, rather than as a theory – is completely subservient to this law, which is a *universal* law and admits of no exceptions. Yet evolution-theory demands that parents of any kind passing on to their descendants what they themselves have *not* got between them has been *continually* happening, every time a species of any kind of living thing has turned into a different species.

Law 6: *The law of biogenesis.* This says that *life can proceed only from life.* Scientists have been constantly trying to create life artificially in laboratories during the last century and more, but never with slightest sign of success.

I call to witness the late Sir Francis Crick, co-responsible with James Dewey Watson for modern gene and DNA theory. Despite styling himself as a sceptic and an agnostic with "a strong inclination towards atheism", Crick found himself forced to teeter on the edge of considering the origin of life miraculous:

> **The origin of life seems almost to be a miracle, so many are the conditions which would have to have been established to get it going.** (*Life Itself*, **Simon & Schuster, 1981; page 88.**)

Law 7: *The fixity of species*; one might even say the *obstinate* fixity of species.

Whenever any new species starts to come into existence, as when two animals of different species manage to be fertile together, nature, far from taking advantage of this, throws the "new" species out within a generation. Thus, invariably, the mule (or hinny) that results from a horse and a donkey mating either (a) is infertile or (b), very rarely, has progeny which revert back to one or other of the original parents. Mules and hinnies themselves produce no further generations of mules or hinnies. The same is true of the hybrid ligers and tigons that are occasionally produced from lions and tigers in zoos (but never in the wild), the zeedonks, the beefaloes, the wolphins and a few others.

Law 8: This item is not technically a law, and, although the relevant law probably exists, I have not yet located it. Rather, it is a statement of undoubted fact. It is that most versions of the theory of evolution depend on mutations which are favourable, and which therefore give a better chance of survival. But in reality, notwithstanding the claims of some scientists, genuine mutations are *never* favourable; *absolutely never*. They are *always* the result of something having gone wrong, and are *al-*

ways more of a threat to existence than vice versa. Even though, in theory, the addition to what is normal might look to be an improvement, the reality is that the addition is a disadvantage, to say the very lease.

Chapter 4

Evidence of design in the operations of nature: the eye; the digestive system; feathers; wing-movement in insects; the honey-bee; instinct; the ratios of the sun, the moon and the earth in relation to each other.

There is yet another problem with natural selection. This is that any change must be of *immediate* value to its possessor if it is to give it a better chance of survival than its rivals. Of what survival value, it must therefore be asked, is, for instance, the first beginnings of an eye? Of what use is a hole in the front of a small globe in the face of any living being that enables light to pass through it, if there are no cells at the back of this globe to receive the light? What is the use of a lens that is capable of forming an image if there is no nervous system capable of interpreting that image? Or was there already such a nervous system? – in which case how could this have evolved before there was an eye capable of giving it information?

Another example: of what survival value is a pair of forelimbs sprouting forth to flap about feebly and recklessly in anticipation of becoming, in a few million years' time, a pair of wings? Natural selection, it must be remembered, is mindless and therefore cannot be expected to decide to tolerate a short-term disadvantage for a long-term gain.

Let us remind ourselves of the solution proposed by Professor Coyne on page 42.

What can be done, he says, is to make educated guesses as to how modern birds evolved by means of natural selection. It is probable, for instance, that early dinosaurs evolved their longer front legs in order to be better able to "grab" their prey and then to deal with it. This "grabbing" would in turn cause muscles to evolve that would make their front legs stretch and also move inwards at the same time, which are the exact movements needed for the downward stroke of a bird's flight. That process would be followed by the birds acquiring the "feathery covering" that they have, which was probably needed for insulation.

What Professor Coyne is inviting us to suppose is that the relatively slow-moving – compared to the wing-beat of even the slowest bird – forelimbs of animals, even the fast-moving animals, were able develop the strength and speed of movement needed to lift the animal off the ground and at same time to acquire the additions to the body of the highest complexity and efficiency that feathers are. This only needs to be pondered over for a few seconds for it to be seen to be evidently ridiculous.

Even Darwin, who, as we are now about to see, only managed to believe his own theory by refusing to allow himself to face the difficulties, at this stage says this with becoming candour, in chapter 6 of his *On the Origin of Species*:

> **Long before having arrived at this part of my work, a crowd of difficulties will have occurred to the reader. Some of them are so grave that I can never reflect on them without being staggered...**
>
> **To suppose that the eye with all its inimitable contrivances for adjusting the focus to different distances, for admitting different amounts of light and for the correction of spherical and chromatic aberration, could have been formed by natural selection, seems, I freely confess, absurd to the highest degree...**

Within two pages, however, he had driven the absurdities out of his mind:

> **If it could be demonstrated that any complex organ existed, which could not possibly have been formed by num-**

erous, successive, slight modifications, my theory would absolutely break down. But I can find no such case.

Nevertheless, absurdities they remain, no matter how determinedly Darwin set about forcing himself to forget them.

Indeed, it is surely worth mentioning that in fact he did *not* manage to forget them. By 1860, the year after publication of his *On The Origin of Species*, Darwin, the almost universally-revered scientist whose genius revolutionised biology, philosophy, religion and man's entire concept of himself, wrote a letter to Asa Gray which included this astonishing passage:

> ...but I grieve to say that I cannot honestly go as far as you do about Design. I am conscious that I am in an utterly hopeless muddle. I cannot think that the world, as we see it, is the result of chance; and yet I cannot look at each separate thing as the result of Design... Again, I say I am, and shall ever remain, in a hopeless muddle. (*Life and Letters of Charles Darwin*, Vol. II. Francis Darwin, editor. D. Appleton and Co., New York, 1899. Page 146: Letter to Asa Gray, 26th November 1860.)

* * * * *

The eye.

Let us now have a really close look at the eye, whether of human beings, animals, fishes, or insects. Here is Professor E. H. Andrews, on page 48 of his book *From Nothing to Nature* (Evangelical Press, U.K., 2000):

> It is obvious that a creature with an eye has advantages over a similar creature without an eye. But what advantage has a creature which has evolved only part of an eye? If the lens of the eye had evolved but not the light-sensitive retina, the creature would be just as blind as it was before. Even if the whole eye had evolved bit by bit, the animal would still be blind until the optic nerve and brain cells had also evolved.

The more carefully and thoroughly we investigate, the more preposterous is what emerges; and although Darwin showed that he was aware of them to a considerable extent, it is hardly possible that he took the trouble to consider all the problems involved in relation to the theory that the eye has somehow evolved. Assuming that Darwin was honest in his beliefs, rather than a purveyor of propaganda which he knew to be untrue -- which, sadly, is something that we cannot justifiably be confident of, as we shall be seeing in chapter 11 – there remains the uncomfortable fact that, if he had forced himself to try to solve the problem that he set himself by referring to the human eye, his book *On the Origin of Species* would surely have died stillborn at chapter 6 and never been submitted for publication.

The following somewhat lengthy but brilliant analysis of the human eye is taken from a valuable book, written in the context of religion, *Apologetics and Catholic Doctrine* by the Rev. Michael Sheehan (M. H. Gill & Son, Dublin, 1950).

Proof of the Existence of God from Order in Nature.

A. Order explained by examples.

(a) The photographic camera.

The photographic camera is a familiar object nowadays. It consists of a small case into which are fitted a sensitive plate and at least one lens. The plate is a little sheet of glass on which is spread a chemical preparation. It is called 'sensitive' or 'sensitised', because it retains any picture made on it by light-rays. The lens is of glass or other transparent substance, and has the power of casting on a screen the image of any object placed in front of it. The camera is completely closed but for a small opening in one of the sides. Through this opening, the light-rays enter: they pass through the lens, and fall on the sensitive plate where they make the picture.

Worth noting in particular are at least the following essentials of a satisfactory camera.

(1) A case blackened within.

(2) A circular opening which can be altered in size so as to admit only the exact amount of light, required.

(3) A lens of a special curved shape.

(4) A sensitive plate.

(5) An arrangement by which the lens can be adjusted to a particular distance from the sensitive plate, so as to secure the proper focus, and to save the picture from being blurred.

All these things were shaped and brought together for the purpose of producing a good picture. We have here an example of order or design, i.e., a combination or arrangement of different things in order to produce a single effect.

(b) The human eye.

The human eye is similar in structure to the camera. Note the following points of resemblance:

(1) The eye-ball corresponds to the case.

(2) The pupil corresponds to the circular opening: it is of adjustable size, and can be altered according to the amount of light required.

(3) The crystalline lens corresponds to the lens of the camera.

(4) The retina corresponds to the sensitive plate.

(5) An arrangement for focussing in the camera, which is done by altering the distance between lens and plate; in the eye by altering the curvature of the crystalline lens.

Here again we have an example of order, because different things are combined to produce a single effect.

Each contributes in its own measure towards the same end, viz., the formation of a clear picture on the retina.

B. Order Demands Intelligence.

How did the camera come to be made? You have your choice of just two answers, viz., that it was made by chance or by intelligence. Now, you know that it could not have been made by chance: such an explanation is so foolish that you would regard it as a jest. You need no help whatever to convince you that the camera was put together by an inteligent workman.

How did the human eye come to be made? By chance, as evolutionists would have us believe? No: that is an absurd reply. The human eye was made by some intelligent being.

C. The Maker of the human eye possesses power and Intelligence without limit.

Make the following suppositions:

– First suppose that all the parts of a camera lay scattered about the table.

– Then suppose that you saw them rise up and move towards one another and fit themselves together.

Would you say that what you saw happened by chance? No; you would say that it was brought about by some intelligent, though invisible, worker, and you would add that this worker must indeed possess very wonderful powers.

Now go a step further. Suppose that the case, the lens and the sensitive plate were all ground to the finest powder and mixed thoroughly together; suppose that the minute fragments of each part sought one another out, and fastened themselves together again; and suppose that each part thus completed took up its proper place so as to give us a perfect camera – would you say that *this* was due to chance? No; and you would protest that here there was need of a worker who was still more intelligent and still more powerful.

I invite you to make one supposition. Let it now be supposed:

first, that you saw just a single tiny speck of dust on the table before you;

secondly, that, having grown to twice its size, it broke up into two particles, and that each of these two particles, having doubled its size, broke up into two others;

thirdly, that this process of growth and division went on, and that, during its progress, the particles managed to build up the case, lens and plate.

Let now be supposed, in other words, that you saw one and the same minuscule fragment of matter produce such widely different things as the camera-case with its blackened sides, the transparent lens with its mathematically accurate curvature, the sensitive plate with its chemical dressing, the aperture with its light-control, and, last of all, the mechanism for focussing. What would you say to such a supposition? You would be tempted at once to denounce it as utterly improbable. You would protest, and with good reason, that only an *all-powerful* being could get a single speck of dust to behave as we have described, to make it multiply itself, and while so doing, to form unerringly, and piece together, an ingenious mechanism.

But is there really any improbability in the occurrence of which we have just spoken? No; the very eyes with which you have been reading this page are witnesses against you. Each of them began as a single particle of matter; and the hidden worker acted upon it, made it multiply itself millions of times and made itself develop such utterly distinct things as the eye-ball, the retina, the crystalline lens with its controlling muscles, the contractile pupil, along with other parts equally marvellous which it is unnecessary to mention. That hidden worker is a being whose power and intelligence our minds cannot measure.

I think it well worth continuing with what Sheehan is saying here, because he certainly makes interesting and useful points.

D. The Maker of the Human Eye Is God.

He who has made the human eye is a spirit; He is a spirit because He is an active intelligent and invisible being. He is one to whom nothing is hard, let alone impossible. We call him God.

E. Further Evidence for This Conclusion

The human eye, as we have explained, grows from a single particle of matter; but the entire body with its flesh, blood, bone, muscle, its various limbs and organs, grows in precisely the same way. It begins as a single living cell which multiplies itself, and gradually forms every part. That living cell, small as it is, is far more wonderful than any machine that man has ever made. You can show how a watch does its work; you can show how the movement of the spring passes from one part to another, until finally it is communicated to the hands. You cannot, however, show how the living cell does its work. It is wrapped round with mystery. Why? Because the mind that made it is too deep for us to fathom.

The mystery lies not only in the *manner* in which the cell works but in the *results* which it produces. As fruit, flowers, foliage, bark, stem and roots come from a single seed, so the wonderful powers of man, his sight, his hearing, his other senses come from the living cell. The more intricate and ingenious a machine is, the greater testimony it is to the cleverness of the maker; but there is no machine in the world that can be compared with the living cell which builds up a man capable himself of making machines and of attaining to eminence in art and science.

* * * * *

The Digestive System.

Let us now, still with Sheehan, look at another example of clear-cut evidence of design in the animal world:

> Another remarkable instance of design appears in the set of organs for the reception, mastication, and digestion of our food. The mouth, with its flexible muscles by which it opens and closes and receives the food; the tongue and the palate register its agreeable or disagreeable taste; the teeth cut and crush it; the salivary glands pour out their juices to prepare it for digestion; the muscles of the throat draw down the masticated food through the alimentary canal to the stomach, where the digestive juices convert it into such a form that it can bring nutrition to every part of the body.
>
> This admirable system of organs, all conducing to the achievement of a single purpose, viz., the preservation and strengthening of life, bears the unmistakable impress of design.
>
> Also, the power displayed in the development of the living cell is on a par with the wisdom that is so evidently displayed. It is a power exerted, not through hands and muscles, but by a mere act of the will.

* * * * *

The reality is that, as soon as evolution is examined at all closely, the impossibilities underlying the theory become endless.

The following are further samples of such impossibilities.

The first two taken from a booklet titled *The Case Against Evolution* by Wallace Johnson, published by the CCCR in Kentucky, U.S.A., in 1970.

Feathers.

Feathers are evolution's worst stumbling block. Feathers are miracles of natural engineering. There is a shaft with dozens of barbs coming from it at a slant. Under a microscope you see hundreds of thousands of little barbules. Then on these there are millions of tiny barbacelles. Then, on many of the barbacelles, the tiniest hooks which, in flight, interlock the whole structure into a plane of wonderful firmness, elasticity and lightness. Then, in an instant, the whole thing can be suddenly unlocked to let the air pass through.

So how did a *reptile* get these wonderful feathers?

Some will claim that the reptile's scales mutated into feathers. Well, that would not be a mutation – it would be a miracle. Others will claim that friction of air did it, friction of air on the scales of generations of reptiles, as they jumped from tree to ground, of reptiles who ran along the ground waving their forelimbs. Friction of air, over vast time, changing scales into feathers.

Well, what do you think? Douglas Dewar, Fellow of the Zoological Society, said: "If this is a serious science, then so is the story of Cinderella."

There is no white pigment in the feathers of a gull as there is black pigment in the feathers of a crow. Pound a crow's feather, crush it up, and all the fragments are still black. Pigment colours them down to the tiniest particle. Smash similarly the chromatically brilliant feather of a humming bird or parrot, and its metallic hues disappear entirely. They are the product, not of pigment, but of tiny grooves or striations set at such an angle that they reflect to our eyes only the wave lengths of light that represent the colours we see. The whiteness of the gull's feather, however, derives from still another source. Neither pigment nor striation explains it. It is due to the infinite number of minute air spaces within the hollow, horny cells. Acting like the bubbles in masses of foam, these air spaces band and reflect the light rays to create the effect of immaculate whiteness that we see.

To produce a particular colour scheme, there may have to be four or five different colours on one feather; each colour must be in exactly the correct position on that particular feather; and that feather must grow in exactly that position on the bird. Now multiply this by thousands and you have some idea of the marvellous miracle which transpires in the clothing of one bird.

Wing Movement in Insects.

If feathers are evolution's worst stumbling block, the means by which insects fly is surely scarcely a lesser one. Continuing with Wallace Johnson:

> Wing movement in insects is complex and consists of elevation and depression, fore and aft movement, pronation and supination (twisting), and changes in shape by folding and buckling. The wingtips describe a figure-8 pattern. Many insects can hover or fly by changing the angle of the figure-8. Some of the very good fliers (*Diptera*, *Hymenoptera*, and some *Lepidoptera*) can fly sideways or rotate about the head or tail by employing unequal wing movement...
>
> The *Hymenoptera* and *Diptera*, and some *Lepidoptera* such as Sphingid moths, must combine excellent flying ability with a relatively small wing area. A honey-bee, for example, could not function well in its hive if it had large wings which are bulky, even when folded over the back, as in the *Papilio* or the Dobson Fly. They compensate for a relatively small wing area with a very rapid wing-beat.
>
> Wing-beat frequencies vary from 55 per second for some beetles, to over 200 per second for the honey-bee, and an incredible 1,046 per second for a midge (*Forc-

ipomya). Clearly, nerve tissue is not capable of firing this many times a second. These insects move their wings by an indirect, asynchronous muscle scheme. Opposing pairs of muscles act to depress and elevate the top of the thorax, to which the wing bases are attached. With a portion of the thorax as a fulcrum, the wings are levered up and down. A single motor nerve stimulus begins a cycle in which the contraction of one member of a muscle pair stretches the opposing muscle and stimulates it to contract.

This process can be repeated several times before another nerve stimulus becomes necessary to reinitiate the process, so very high wingbeat frequencies can be obtained.

Notwithstanding the detailed information that Wallace Johnson has just put in front of us, the complexity of all the different mechanical systems, all having to work perfectly together, is actually beyond our comprehension. Yet we are asked to believe that it developed bit by bit over millions of years by a succession of *chance* happenings?

* * * * *

The Honey-Bee.

It is in fact doubtful whether evolution's worst stumbling block is either feathers or insects' wings. It must surely be the just-introduced honey-bee. Still with Wallace Johnson:

> How does it come about that a thousand different hives in various parts of the world, having no communication with each other, all work to the same accurate scale in exactly the same way as engineers all work to a standard measure? So accurately do the bees work that if you were to compass a section of comb from each of these 1,000 hives, there would not be the slightest variation. The most skilful draughtsman could not reproduce this even on paper without special instruments. How does the bee accomplish this, and who finalised the design and set the scale?

The barbs on the bee's sting are designed to make their extraction difficult. Seeing that this scientific device costs the bee his life in exchange for one injection, it is reasonable to suppose that this is either spontaneous adaptation or the bee's own design. This is especially significant when we examine his relative the wasp. The wasp is a far less intelligent creature yet he has a more deadly instrument than the bee, and with it he can administer repeated doses of the poison without harm to himself.

In this little insect, the honey-bee, we have a trap to catch and baffle the ablest men that ever tried to support the evolutionary theory. In it we have a highly endowed little creature with instincts that seem to rival reasoning powers more closely than the instincts of any other creature, and yet... there is no door left open for the entrance or the transmission of these wonderful peculiarities. The parents of the bee – the queen and the drone – have none of these instruments to transmit. The honey-bee itself – which has no offspring – can transmit nothing.

Darwin's theory is closed at both ends.

*   *   *   *   *

Instinct.

Another impossible problem for those who wish their belief in evolution to be rationally based is set by the various *instincts* that exist in the animal world. Here are three examples of such instincts in action, similar in that they all involve flying.

The so-called wild rock pigeon is often referred to as a homing pigeon since it has an astonishing "in-built" homing ability. If it is caught where it lives and then released anywhere in the world, normally it will somehow find its way back to its nest. With such birds as have been specially bred for racing, flights of well over a thousand miles at speeds of about sixty miles an hour (and top speeds of about a hundred miles an hour) have been recorded. *How* could this inherited capability, of doing what we can do only with the help of sophisticated instruments, have evolved?

A second example, even more remarkable, is the Arctic tern. Born in the Arctic circle, this little bird, a mere four ounces (a hundred grammes) in weight, migrates down towards the opposite pole, the *Antarctic* circle, and then later in the year returns to its nesting area in the Arctic. The "round trip" with its various detours may be 44,000 miles, and it does this every year for its life-span of, commonly, about thirty years. In total, the Arctic terns can cover a *staggering* million and a half miles -- equal to three trips to the moon and back.

Let us make the assumption, however improbable, that a group of birds residing near the North Pole would have improved the chances of survival of those birds by looking for an area better suited to the non-nesting portion of their lives, rather than by learning to adapt to their existing surroundings throughout the year. Let us then make the further assumption that they therefore launched themselves from time to time into the unknown.

How far would they have travelled without getting lost? What were they looking for?

Supposing, however absurdly, that they had in fact known what they were looking for -- and let us not forget that it is not safety for their young that is on their mind, because they are leaving their nesting-home -- how could they possibly have developed the ability to fly five thousand miles across the Indian Ocean without stopping once? Had they developed the skill of sleeping while flying before they attempted this flight, or did this skill develop bit by bit, with all those attempting it perishing, millennium after millennium, until one finally, purely by chance, "made it"?

Apart from anything else, was there no problem in coping with the extremes of temperature they would have met on the way? And how many of them perished before one of them mastered astronomy and navigation and passed this newly-acquired learning on to its descendants?

Highly complex astronomy and navigation at that. Bear in mind:

(a) that the view of the night sky changes with the seasons and with the locality of the viewer; and

(b) that our adventurous tern needs an accurate method of measuring both date and time, to enable it to compare the night sky that it sees at any moment with the appearance that the night sky ought to have if it were on its correct course; and that it also needs the capability of calculating a suitable correction.

Most migrating birds indicate a very much greater knowledge of navigation than we have if we are deprived of artificial instruments. Certainly it is the case that, when the sun or the stars are hidden by clouds, the bird finds its journey to its destination a more difficult and a more erratic process, thereby showing, incidentally, that it does actually use the stars and sun for navigation. Nevertheless it does not, as we human beings would, lose its way and go round in circles or in the reverse direction; and this shows that it possesses a few navigational aids that we, with our, to say the least, superior intelligence, have so far been unable to manufacture.

Further, having somehow acquired its knowledge of astronomy and navigation as it was going along, how did this bird pass this remarkable discovery on to its chicks and other members of its species? Did these thus-informed creatures remember all this the first time that they were told by their parents? Or did they all perish so that the whole process had to start all over again?

Finally, the stage has been reached, in our present day, where many species of birds have become so skilled at educating their young in the necessary arts of time-telling, geography, astronomy, navigation, and so on, and so confident in the success of their teaching, that the parent bird invariably migrates before its chick does – weeks earlier in some cases. We may almost suppose that, having adopted an impossibly difficult course of action, they then set out to make it even more difficult.

Really, it will not do. The migration of birds, far from being capable of *creating* a new species, can do no more *eliminate* the *weaker* members and *preserve* the *existing* species.

Could anything be more impossible to explain by any possible variation of evolution-theory than those two migrating birds, the homing pigeon and the Arctic tern? Arguably, yes: at least one further example of migrating creatures in fact could. This is the monarch butterfly of North America, measuring about four inches across when its wings are outstretched and weighing about one fifty-sixth of an ounce (half a gramme), surely a strong candidate for our recognition as the most remarkable of all living beings on earth.

Some of these butterflies migrate each year on a journey of *some three thousand miles*, starting in southern Canada in September, then flying down through the United States to central Mexico, and beginning their return to Canada in March of the following year. Perhaps more remarkable than anything relating to migrating birds, scarcely any of these butterflies completes its journey. Most of them pause, lay eggs, and then die, leaving their offspring to continue the journey. This means, of course, that they are unable in any way to pass on to their offspring what it is that they are engaged in and how to accomplish it.

The same sequence of events takes place on the journey back to Canada.

How do they reach their destination, those that do? Scientists have done as much research as they can, but to little effect. At first it was thought that these butterflies used the position of the sun to work out what direction they should be travelling in at any point in the journey. That cannot come within any distance of being a complete answer, however. It does not explain how they are able to continue to proceed in the right direction on cloudy days.

When the solution of their making use of the sun to keep going in the right direction had to be ruled out by those doing the research, it was supposed as a possibility that their brains, if that term can be used in the context of butterflies, were sensitive to the magnetic field of the earth, as a compass is, and that they must be capable of using this faculty as well as being guided by the position of the sun.

Either way, it is completely impossible to see how this astonishing navigating ability could have gradually *evolved*. To say the least, every difficulty faced by evolutionists in any attempt to explain how evolution of birds could have come about applies *even more so* to any supposed evolution of monarch butterflies.

Monarch butterflies, incidentally, are by no means the only species of butterfly that migrates. To take two more examples, the painted lady butterfly and the hummingbird hawk moth: both of them undertake migrations of almost a thousand miles between Britain and North Africa. (Much of the foregoing information was taken from an article, written by Richard Gray, the *Sunday Telegraph*'s Science Correspondent, on page 8 of that newspaper on 27th June 2010.)

The reality is that what instinct does is to enable animals, birds and even insects to act intelligently even though they do not have intelligence. In some cases, indeed, it enables them to act even more intelligently than we ourselves can act. How did spiders first reason out the principles of geometry without which a spider's web could not be constructed? How are these principles passed unerringly from spider to spider without instruction or even contact, while educated human beings find it by no means easy to pass on *much* simpler principles to their children even with the help of years of education?

And all these astounding capabilities of insects, fish, birds and animals exist even though...

Turning once again to Byron Nelson's *After Its Kind*, on page 146:

No animal, however, has the capacity to reason, by which is meant the capacity to handle abstract ideas. As Professor F. O. Jenkins says:

"What dog or ape that warms himself by the fire, and has seen good wood put onto it time and again, ever has sense enough to bring sticks of wood to it himself when he sees it dying out and feels himself getting cold?"

* * * * *

Here is another important fact to take into account. On the, I submit now-discredited, hypothesis that the theory of evolution is in fact a *valid* theory, the main principle underlying it – the principle that everything happened at random and by chance, with "the fittest" surviving to continue the evolution-process – must apply equally to much else in the universe. Specifically, it would have to apply as well to purely physical non-living things wherever there is clear order in composition and arrangement. Let us look at just one example.

The Ratios of the Sun, the Moon and the Earth in relation to each other.

In all the foregoing in this chapter we have concentrated on giving clear evidence of design as a necessary explanation of various species of *living* beings. For the sake of reasonable completeness in examining this interesting subject, it is worth supplementing that with a piece of evidence, equally clear, that does *not* involve living organisms. I have chosen for this purpose one that I believe to be the most spectacular of all: the relationship of the sun and the moon, in (a) their respective sizes and (b) their respective distances from the earth.

Specifically, (a) the sun is four hundred times *bigger than* the moon, and (b) it is *also* four hundred times *further from the earth* than is the moon.

Using a technical mathematical term, the angular diameter of the *sun* as seen from the earth and the angular diameter of the *moon* as seen from the earth are therefore identical at 0.5°. Because of this, the moon is capable sometimes of blotting out the sun *exactly* in an eclipse – *in their appearance*, neither of them is either bigger or smaller than the other.

In other words, the effect, viewed from the earth, is of two circles of identical size superimposed one on top of the other.

The odds against this occurring by accident are incalculable.

Chapter 5
Evidence from Fossils.

I have brought up the subject of evolution with many people over the years, and the common thread that I have met in response to what I have said has tended to be:

"Fossils by themselves amount to sufficient proof that evolution took place."

And, as we saw in chapter 3 of PART II, Professor Coyne made extensive use of fossil evidence to support his case. For a further look at whether such proof does in fact exist, I turn back to the book by Roger Gallop, *The Greatest Deception in Modern History*, that I made use of in chapter 2. From its chapter, "Lack of in-between types":

If evolution were true, there should be billions of transitional forms. Or to put it another way, if there were a slow evolutionary transition from amoeba to higher amoeba to man over billions of years, one would expect to find billions of examples of transitional fossils and living in-between forms. For instance, if the limb of an amphibian had slowly transitioned into the wing of a bird, one would expect to find a long succession of fossil forms with multiple stages of transition over these billions of years – one stage gradually transitioning to another. So where are all the in-between forms?

Darwin had admitted that the absence of in-between stages was "the most obvious and gravest objection which can be urged against my theory." And evidence is still not forthcoming. Evolutionist Dr. Colin Patterson of the British Museum of Natural History had this to say, when he had failed to include illustrations of transitional form in a book he had written on evolution:

"You say that I should at least show a photo of the fossil from which each type of organism was derived. I will lay it on the line – there is not one such fossil for which one could make a watertight argument."

Gallop continues:

> Where is the chain of evidence? Why isn't every geologic stratum full of intermediate form? In fact, *there are none.*

Not a single unequivocal transitional form with transitional structures in the process of evolving has ever been observed within the billions of known fossils. Each basic kind of animal is distinct in our modern world and in the fossil record, although there is much variation within these basic groups or kinds. Variation (speciation and natural selection) within animal kinds is not evolution.

In his book *Sudden Origins* (John Wiley & Sons, April 1999), Jeffrey H. Schwartz says on page 300:

> **Instead of filling in the gaps in the fossil record with so-called missing links, most paleontologists [have] found themselves facing a situation in which there were only gaps in the fossil record, with no evidence of transformational intermediates between documented fossil species.**

To that, Dr. Henry Morris, in an article, "The Scientific Case against Evolution" in the periodical *Acts and Facts* of December 2000, adds:

> **The entire history of evolution – from the evolution of life from non-life to the evolution of vertebrates from invertebrates to the evolution of man from the ape -- is strikingly devoid of intermediates. The links are all missing in the fossil record, just as they are in the present world.**

* * * * *

Under the heading of fossils, as we saw in chapter 4 of PART II, Professor Jerry Coyne leans heavily on *Archaeopteryx*. If there is indeed a third species in the fossil record as well as dinosaurs and birds, it does not follow that this was transitional, rather than another species on its own that has become extinct. It might of course be argued against this that the rock in which it was found is identifiable as to its approximate date. The dating of rocks, however, will be shown in chapter 7 to be unreliable, to say the least.

Chapter 6
More on Fossils.

Now let us move on to what are claimed by evolutionists to be fossils of the common ancestors of the species of ape and the species of human being.

I am going to start with Piltdown Man (*Eoanthropus*) although as we have already seen this has been exposed as a fraud. Once again, I am making use of Roger Gallop's valuable book, *The Greatest Deception in Modern History*.

This fossil was found in a gravel pit in Sussex, England, in 1912 and was considered to be the second most important fossil proving the evolution of man. For over forty years it was widely publicised as the "missing link" but in 1953 it made newspaper headlines that it was a hoax based on a human skull cap of modern age and an orang-utan's jaw. The fragments had been chemically stained to appear old and the teeth had been filed down. How it could take four decades of examination before the truth came out beggars belief. It rather points to the idea that more than one evolutionist did not want it exposed as it had been such useful evidence for their theory.

Other frauds include:

The "Nebraska Man" fossil that was based on a single tooth in 1922. Nevertheless, this fossil tooth grew to be considered an evolutionary link between man and monkey. In 1928, it was discovered that the tooth belonged to "an extinct pig". What is more, in 1972 a herd of these pigs, *still existing*, was found Chaco, a region in Paraguay.

"Ida" (*Darwinius masillae*), named in honour of Darwin and recently promoted by the media as the "missing link", is nothing more than a fossilised lemur – yet another false claim that has been quietly discarded.

"Neanderthal Man" (*Homo sapiens neanderthalensis*) was named as an ape-man after fossils were found in Germany in 1856. Depicted as "brutish and stooped" these "men" were supposed to have dragged their knuckles along the ground, and so were hailed as the perfect "in between"; but now many scientists admit that they are humans, just like the rest of us, and that their stooped posture was caused by the disease called rickets.

"Lucy", who needs to be considered at some length because Professor Coyne uses her as what he considers to be important evidence.

I am taking what follows from a Mr. Bill Nugent, who some time ago posted it as an article on the Internet. It is lengthy but, in my judgement, valuable enough to be well worth including.

On November 24, 1974, American anthropologist Donald Johanson and a graduate student were making a routine search for fossils in a gully in the Afar Depression in Ethiopia when a small bone fragment caught Johanson's eye. They dug around and found several more fragments of bone with no duplications and this indicated that the many fragments were from one individual.

Excitedly, they returned to their camp and announced the initial findings. Someone in the camp was playing the Beatles' song *Lucy in the Sky with Diamonds* repeatedly during the evening. They decided on the name "Lucy" for their soon-to-be-famous find from that song. The entire team returned to the gully the next day and unearthed bone fragments that were the remains of about 40% of the skeleton of what even secular evolutionist researchers describe as similar to a small chimpanzee. They called the new species of extinct ape *Australopithecus-Afarensis*, which means *"Southern ape of the Afar region."* The fragments were removed to the Cleveland Museum of Natural History and assembled and studied. There were intense debates among evolutionary paleontologists over this unusual find. All sides agreed the skull and brain case are exceedingly small and must have housed a tiny undeveloped apelike brain. The evolutionists also agreed that her upper extremities appear similar to those of modern tree dwelling primates.

The real debate, the real significance, was the pelvis and leg bones that are of such a structure that some evolutionists contend that Lucy would have been bipedal, which is to say that she could have walked upright much of the time. But this claim of bipedalism depends on how they assemble the fragments of pelvis and leg bones.

Some of the evolutionists have noted that the orientation of the iliac blade of the pelvis matches that of chimps, not humans. This clearly undermines the claim that Lucy was bipedal. Evolutionist Christine Berges used high tech analysis techniques and came to the conclusion that Lucy's bipedalism differed from that of humans.

All of us have seen chimps walk on their hind legs and notice how they sway from side to side as they walk. This is because the chimp pelvis is structured differently from that of humans. The claim of bipedalism is far from any degree of certainty.

There is actually very little evidence in Lucy from which scientists make their bold assertions. Remember, this is just a partial skeleton and there are, of course, no soft-tissue remains. It is educated men making educated guesses. They view the evidence through the lens of their Darwinian presuppositions. More evidence has piled up over the years as more *Australopithecine* fossils have been discovered in other parts of Africa. The evidence suggests that Lucy and all those of her species were fully ape.

The other, more recent finds of *Australopithecine* fossils clearly undermine Lucy from being considered a human ancestor. The toe bones are curved like those of tree-dwelling apes. *Australopithecine* wrist-bones show the ability to lock to enable wrist walking. Evolutionists Brian Richmond and David Strait showed how the neck, shoulder, arm and finger anatomy was suited for tree dwelling. This is not a case of *our* scientists being able to beat up *their* scientists. Their scientists are furnishing the evidence to destroy Lucy. Lucy was just an ape.

The Christian creationist group *Answers in Genesis* has a Lucy display at their Creation Museum in Ohio. It shows a model of how Lucy would have appeared and shows her on all fours as a chimpanzee-like animal. This model is based partly on the work of Dr. David Menton. Dr. Menton earned a Ph.D. in cell biology from Brown University. He also served as an associate professor of anatomy at Washington University School of Medicine (St. Louis). Dr. Menton is also featured in a video titled, *Lucy -- She's No Lady*, that demolishes any notion that Lucy walked humanly upright, and he clearly explains why Lucy was fully ape. Ape to human evolution is nothing more than pseudo-scientific fantasy!

All of the discoveries of bone fragments that are splashed across the media from time to time with claims of "missing link" are either fully ape or fully human. *Australopithecine* are fully ape, Neanderthal Man is fully human.

* * * * *

Before we leave the important and controversial subject of fossils, I should like to call upon one last witness on this subject, Ernst Haeckel of Germany (1834 –1919), the leading promoter of Darwin's work outside Britain.

On the face of it, he is indeed as important a witness as any reader could reasonably want. Almost unbelievably, in the number of subjects in which he was a recognised expert, he was a zoologist, a naturalist, a philosopher, a physician, an academic professor, a marine biologist, and even an artist; and he discovered, described and named thousands of new species, mapped a genealogical tree relating all life forms, and coined many terms in biology, including ecology, phylum, phylogeny, and Protista. Furthermore, his published artwork includes over a hundred detailed, multi-colour illustrations of animals and sea creatures, collected in his *Kunstformen der Natur* ("Art Forms of Nature").

Of particular interest for our purpose here: under the heading of philosophy he was the author of *Die Welträthsel*, in English *The Riddle of the Universe*, written for the express purpose of supporting Darwin's theory of evolution, translated into scores of languages and a world-wide best-seller to the extent of many millions of copies over the years, a remarkable achievement for a book devoted to science, although of course one shared by Darwin's *Origin*. In fact, in Germany it is Haeckel and not Darwin who is the popular apostle of "Darwinism".

Although Haeckel was originally revered, and of course very understandably, it eventually emerged that he was a scientific falsifier and a disgrace to his many professions.

In what now follows, I am largely quoting from a book called *False Prophets* (published in 1929) by the Rev. J. M. Gillis.

In *The Riddle of the Universe*, Haeckel writes:

"In the last twenty years a considerable number of well-preserved fossil skeletons of anthropoid and other apes have been discovered, and amongst them are all the important intermediate forms which constitute a series of ancestors connecting the oldest anthropoid ape with man."

If this were true, the "missing link" would no longer be missing. But Alfred Russel Wallace was writing, at about the same time: "There is not merely one missing link, but at least a score of them." And Father Muckermann declares:

> "Haeckel's curious '*Progonotaxis*', or genealogy of man, is pure fiction. It consists of thirty stages, beginning with the 'moners' and ending with *homo loquax*. The first fifteen stages have no fossil representation."

Perhaps even more startling than Haeckel's dishonesty in fabricating drawings, is the fact that he ultimately admitted the fraud. He confessed:

> "Six or eight per cent of my drawings of embryos are really falsified. We are obliged to fill the vacancies with hypotheses."

And now, gracious readers, please give your full attention to what Haeckel says next. Continuing Gillis's account:

> But immediately he declared, in his own defence, *that it is customary for scientists to make use of fraudulent designs.* "I have the satisfaction," he says, "of knowing that side by side with me, in the prisoner's dock, stand hundreds of fellow culprits, many of them among the most esteemed biologists. The majority of figures, morphological, anatomical, histological, which are circulated and valued in students' manuals, and in reviews and works of biology deserve in the same degree the charge of being falsified. *None* of them is exact. All of them are more or less adapted, schematised, reconstructed."

This is interesting, if true. Obviously we cannot be certain that it is true purely on Haeckel's authority. A liar will lie about men as well as embryos. On the other hand, however, a liar sometimes tells the truth. A criminal, undergoing the "third degree" will lie, and will continue to lie for hours or even days. But if he is suddenly cornered and admits one lie it is a well-recognised psychological fact that he will probably break down and tell all the truth he knows.

Indeed, he may tell so many true things as not only to incriminate his companions, but to embarrass his investigators. I suspect that it is so in Haeckel's case. Angered at being caught in a fraud, he "peaches" on his fellow scientists. He admits the lie and then proceeds to "spill the beans".

To give just an example or two of his scientific method. He is enumerating his famous thirty stages in the evolution of man, "fifteen of which have no fossil representatives," when he says:

> "The vertebrate ancestor, number fifteen, akin to the salamanders, must have been a species of lizard. There remains to us no fossil relic of this animal. In no respect did it resemble any form actually existing. Nevertheless comparative anatomy and ontogeny authorise us in affirming that it once existed. We will call this animal 'Protamnion.'"

This, be it remembered, is *science*, not poetry. But was there ever a better example of Shakespeare's assertion that "Imagination bodies forth the form of things unknown... turns them to shapes, and gives to airy nothing a local habitation and a name!"

Yes, Haeckel, wishing to trace the genealogy of *Homo Sapiens* back to the anthropoid ape, declares *a priori* that there were thirty stages, admits that the first fifteen are missing, but assumes them and names them. *And this is science!*

This is quite on a par with the procedure of every scientist who claims to be able to trace the genealogy of man back to the ape. As Alfred Russel Wallace says: in that genealogy there is, not one, but at least a score of "missing links".

Chapter 7
Dating of the universe and its contents, by means of, for instance, carbon-14, fossils, and radio-metric dating.

Of course evolution, if it could happen at all, certainly could not happen if there was insufficient time for the very gradual changes that must have taken place for a single-celled piece of life to have developed into the living beings of every kind that we see all around us. Hand in hand with the theory of evolution, therefore, necessarily goes today's accepted view of the age of the world, which is yet another belief based on assumptions that can be shown to be not merely erroneous but, in some cases, clearly fraudulent as well.

The first such assumption is that pre-history is a suitable subject for examination by science or that science can make a pronouncement of value on events which have not been observed and recorded. It is in fact impossible to extrapolate backwards from what can be observed in the present without running the risk of overlooking a sudden catastrophe, or a supernatural intervention, let alone running the risk of having insufficient knowledge.

Even the tree-ring dating method, long held to be as reliable as a well-made clock, is flawed. From *The Exodus Problem and its Ramifications* by Donovan A. Courville (Crest-Challenge Books, Ca., 1971), volume 2, page 35:

> Recent finds indicate that not even this method can be considered as yielding data usable for the establishment of dates. It is now recognised that under proper conditions a tree may produce more than a single ring in a year, and that three or four rings in a year are not uncommon. Trees growing on a slope where the water runs off rapidly to give a rep-

etition of wet and dry periods may show a multiplicity of rings in a single year; even two sides of the same tree may reveal different numbers of rings.

Every chemical dating-method, such as that involving carbon-14 (produced by cosmic rays high up in the atmosphere), involves radiometric dating. This is based on the fact that some elements decay over a period of time to form new elements; as uranium and thorium, for instance, decay to form lead.

Radiometric dating is based on three assumptions that can none of them be proved:

that the present state of all the lead (or whatever the element may be) is the result of decay;

that the decay rate has remained constant;

and that decay began at time zero.

If "scientific" means accepting as truth only that which can be *scientifically demonstrated*, the decay systems for calculating the duration of time are not scientific. The same scientists entering a room and seeing a two-inch candle burning would not attempt to calculate how long it had been alight unless they knew: the original length of the candle; at what rate it had been burning from the moment that it had been lit; whether the rate of burning was constant; and whether the candle had been continuously alight. Evidently, however, such necessary scientific rigour is not required in present-day palaeontology (the study of extinct beings).

How was the geological "timetable" calculated and put together by the scientists? Good readers, it was *made up*, in the nineteenth century.

It was made up *by guesswork* in order to allow for the huge time-spans demanded by evolution. Sir Charles Lyell, the founder of uniformitarian geology, as opposed to the "catastrophical" geology that had been generally accepted up till then, was the first to take this route, and the guessing has continued ever since. A continual adjustment of the "timetable" has been needed, and scarcely a week goes by without an announcement in the Press that a new geological discovery has forced geologists to amend the official "timetable" yet again.

This timetable is based on what are called index fossils, with the simplest fossils at the bottom and the most complex at the top. Even leaving aside the fact that, as will become clear in chapter 13, dead bodies, when exposed, *decay and disintegrate*, rather than fossilise, the table produces impossible situations. The *top rocks* of the Matterhorn in Switzerland, for instance, are revealed by this dating technique to be hundreds of thousands of years *older* than the *base rocks* of the Matterhorn. This may not be offensive to scientists – and I see no reason to withdraw that sarcasm or to apologise for it – but it is certainly offensive to common sense.

Radiometric dating was developed in the 1950s and has produced equivalent anomalies. Volcanic rocks taken from late eighteenth-century eruptions, and therefore *known* to have been formed within the last two hundred years, have, for instance, been "proved" – yes, supposedly *proved* – by radiometric dating to be 16 million years old. The more accurate carbon-14 dating -- which, even so, gives excessive age because of the diminishing magnetic field, and is also unreliable for the reason given by Dr. Barnes at the end of the next chapter (in subsection (b) of the paragraph there starting "In the same book…") -- has shown the oldest specimen of *homo sapiens*, dated between 100,000 and 200,000 years ago by other methods, to be not older than 8,500 years.

Important in this context are basalt rocks that have been found in the sea surrounding the islands of Hawaii in the Pacific Ocean. Let us remind ourselves of what Professor Coyne has told us, as I quoted him saying back in chapter 3 of PART II:

> **When radiometric dates can be checked against dates from the historical record, as with the carbon-14 dating**

method, they invariably agree. It is radiometric dating of meteorites that tells us that the Earth and the solar system are 44.6 billion years old.

What Professor Coyne states there turns out not to be true. Under scientific testing those rocks were given an age of about twenty million years, but it is a *historical* fact that they were in fact formed by volcanoes *less than two centuries ago*. There are, incidentally, volcanoes there that are still active and still forming islands.

Something else that must raise questions is the dating of coal that modern science has come up with. According to the World Coal Association, the process responsible for the formation of coal began three hundred million years ago, and it would presumably therefore be absolutely impossible for any human artifacts to be found within this ancient substance. Incredibly, however, many items *have*, we are told, been found in such deposits, buried inside inside the coal itself or deep down within coal veins found in mines that have been tunneled out far beneath the Earth's surface.

To give just one example: an iron pot was allegedly found inside a large piece of coal in Oklahoma back in 1912. On this, Frank J. Kenwood made the following affidavit (published by Wilbert H. Rusch in *The Creation Research Society Quarterly 7, 1971*):

> While I was working in the Municipal Electric Plant in Thomas, Okla, in 1912, I came upon a sold chink of coal which was too large to use. I broke it with a sledge hammer. This iron pot fell from the center, leaving in the piece of coal the impression or mould of the pot... I traced the source of the coal, and found that it came from the mines in Wilburton, Oklahoma.
>
> According to Robert O. Fay of the Oklahoma Geological Survey, the coal in the Wilburton mine is about 312 million

years old, which is problematic, to say the least, since *no one* would claim that human beings existed as long ago as that!

There have been hundreds of other finds which have baffled scientists and archaeologists alike -- from a handbell which still rings, to a wall of polished blocks, to even the fossilised imprint of a shoe on which, around its outline, clearly visible stitching with well-defined thread can be seen. All these things would have to be at least as old as the coal, which is obviously nonsensical if the coal is several hundred million years old.

Chapter 8

On the relevance of the earth's magnetic field.

We now turn to a method of measuring the age of the earth that

(a) is never quoted by those who believe in a billions-of-years-old universe,

and

(b) demonstrates *categorically* that the earth cannot be even ten thousand years old.

This is the earth's magnetic field.

At first sight it appears to require assumptions similar to those of radiometric dating. At second sight, however, it can be seen as very much more difficult to reject.

The earth's magnetic field has been measured continuously since 1835. During that time, it has decayed at a rate at which it loses half its life every 1400 years. It is therefore vanishing fairly rapidly. Quoting on this subject what Thomas G. Barnes D.Sc. has to say on page 25 in his book *Origin and Destiny of the Earth's Magnetic Field*, published in 1971:

> If the initial value of the earth's magnetic field were known, a date could be established for the origin of the magnetic field. This would be done by extrapolating backward on the exponential decay curve.
>
> But we do not know what the initial value was...
>
> The magnetic field for a magnetic star (with its thermonuclear source) is about 100 gauss. One would certainly not expect a planet to have a magnetic field as great as

that of a magnetic star. So it is not likely that the earth's magnetic field was ever that great...

Applying the reasonable premise that this planet never had a magnetic field as great as that, one can calculate that the origin of the earth's magnetic field had to be more recent than 8,000 B.C. That is to say, the origin of the earth's magnetic field was less than 10,000 years ago. Just how much more recent than 10,000 years cannot be determined from present scientific knowledge. If one assumes that the initial value of the earth's magnetic field were about one order-of-magnitude less than that of a magnetic star, the origin would have been about six or seven thousand years ago...

The only alternative to a young age for the earth's magnetic field is to deny the existence of decay in the earth's magnetic field, not a very astute stand for a scientist in view of the strong physics on which Sir Horace Lamb's theory is based and the 130 years of world-wide real-time data to substantiate it.

In the same book Dr. Barnes also demonstrates:

(a) that at even as recently as only twenty thousand years ago the heat generated by the magnetic field would have been so great that the earth could not have existed;

(b) that the stronger the magnetic field, the greater the barrier against cosmic radiation, rendering invalid any calculations that are based on the carbon-14 dating method.

Chapter 9

From hydrogen to human?

Back in chapter 2 of PART I, a succinct definition of the theory of evolution as understood by today's scientists was given, courtesy of Professor Coyne, which I believe it to be worth repeating here:

> What the theory of evolution holds is that life on earth began as just one tiny primitive species that suddenly came into existence by chance more than three and a half billion years ago, as something with life in it, and then, over a long period of time, branched out into other species which have continually increased in number.

Here, now, is another summary of the theory of evolution representing the beliefs of modern scientists, an even briefer one than that given by Professor Coyne but just as accurate:

> **Hydrogen is a colourless, odourless gas which, if given enough time, turns into people.**

That comes from page 19 of a book published in 1979, *Evolution: Its Collapse in View* by Henry Hiebert, an author who, as we shall now see, did not believe the theory of evolution to represent reality. His next words:

> After you get through laughing at this hilarious statement, you must come to the realisation that it is the only alternative to special creation, and that it is the exact position, reduced to its lowest terms, that the evolutionists express.
>
> Scientists believe that the primordial stuff of the universe was hydrogen. From hydrogen all things evolved. The agency involved is believed by scientists to have been pure chance! The proof cited by scientists for the astounding conclusion? It is the existence of hydrogen

on the one hand, and the existence of man on the other! Where else, they reason, could man have come from? The only alternative, creation by an omniscient, omnipotent God, is described as "unthinkable" by British anatomist Sir Arthur Keith, and "incredible" by noted evolutionary protagonist Julian Huxley. These scientists thus profess to have some absolutely certain proof that God does not exist.

How Hiebert worded that opening one-sentence summary of his is of course not as Professor Coyne worded his, and doubtless is not how any of today's respected experts in the world of science today would choose to word it. Hiebert does, not exaggerate there, however -- not even in the smallest degree -- what evolutionists are all forced to believe by their acceptance of the theory in general. And people certainly do seem to have to force themselves to believe it. A typical example of this is found in the following letter, reproduced in the periodical *The Week* of 20th April 2019 after having appeared a few days before in the *Times* newspaper, in response to an earlier letter.

> Tom Whipple [an earlier correspondent in *The Week*] says that astronomers have produced "an image of one of the most mysterious phenomena in our universe". An equally mysterious phenomenon is the fact that land creatures with brains capable of producing this astonishing image eventually evolved from one-cell water creatures.
>
> Turid Houston, Ashtead, Surrey.

* * * * *

The article on Evolution in *The New Encyclopaedia Britannica* (1975) includes the following words:

> It is not unreasonable to suppose that life originated in a watery "soup" of pre-biological organic compounds and that living organisms arose later by surrounding quantities of

these compounds by membranes that made them into "cells".

This is usually considered the starting point of organic ("Darwinian") evolution.

Preferring, surely not unreasonably, to avoid speaking in my own name when that is practical, I turn to the author E. F. Schumacher in his book *A Guide for the Perplexed,* originally published in 1971 and republished by Vintage in 1995. On page 127 of the later edition:

> One can just see it, can't one? Organic compounds getting together and surrounding themselves by membranes (nothing could be simpler for these clever compounds) and lo! – there is the cell. And once the cell has been born there is nothing to stop the emergence of Shakespeare, although it will obviously take a bit of time. There is therefore no need to speak of miracles – or to admit any lack of knowledge. It is one of the great paradoxes of our age that people claiming the proud title of "scientist" dare to offer such undisciplined and reckless speculations as contributions to scientific knowledge – and that they get away with it!

When considering the ludicrous speculations by scientists about the origin of life, such as that contained in the *New Encyclopaedia Britannica*, as quoted above, it should also be remembered that the origin of life and matter are not even suitable subjects for science to study. Science studies the world *as it is today*. What science can *not* do is work out what existed before life or the world or the universe began. In the words of Professor E. H. Andrews on page 94 of his book *From Nothing to Nature* (Evangelical Press, U.K., 2000):

> Suppose no one had even seen a caterpillar, but only butterflies and moths. No amount of study of the butterfly would ever show how the lovely winged creature came from a crawling caterpillar.

Nor does the problem end there. Even more fundamentally: suposing it were possible to demonstrate that the universe was made up of countless individual atoms – or indeed of the component parts of these atoms – which by a wonderful series of happy chances had come together to form the totality of what the universe and its contents now consist of, how was each individual component of each atom made in the first place? They could not have made themselves; for, once again, nothing can do any creating until it already exists.

What I believe I can justifiably assert is that people who believe in evolution do so for one of the following reasons, each of which is almost self-evidently an insufficient reason for believing in evolution:

either no-one has suggested to them that the theory is false;

or, they are in a habitual state of blind acceptance of the theory because of the virtually universal consent – and therefore to them authoritative consent – of the scientific world to the theory, and have seen no reason to give the matter independent consideration;

or, having received the suggestion that the theory is false, they have dismissed that suggestion at once, as not worthy of consideration;

or, having been pressed to give the matter consideration, they have obstinately refused to do so;

or, having been forced to give the matter consideration, they have, consciously or otherwise, wilfully refused to believe what is capable of being irrefutably proved by evidence, facts and reason.

The second of those possible reasons – a habitual state of blind acceptance for the reason given above – is perhaps the one most deserving of sympathy. We ordinary folk, if I may so refer to my readers as well as to me, might well, and surely very understandably, feel overwhelmed by what the scientific world keeps telling the rest of us. We also, no less understandably, might feel that, as non-scientists, we hopelessly lack the competence that is needed to give this important matter independent consideration.

This problem will be looked at with some care in chapter 13. Meanwhile, I think we have seen sufficient to make it completely reasonable for me to assert resolutely that it is a clear *reality* that the theory of evolution is scientifically impossible – yes, *scientifically* impossible. It cannot be defended by coherent or reasoned argument. It can *only* be defended by one or more of the methods used in every other fraudulent science: by bluster and ridicule; by the mere weight of the authority of the important people who support the theory; by ignoring the arguments against it; by suppression of the arguments against it; and by sheer weight of constant and continual propaganda.

Chapter 10

Then why was the theory of evolution introduced?
And why has it survived?

Evolutionary theory is so impossibly improbable that the non-science that it is in reality could not have acquired general acceptance by accident. The world at large is the victim of a carefully planned deception. Why? What is the purpose behind the evolutionary hoax?

That is a question that one of the founders of anti-religious Communism, Friedrich Engels, once gave an answer to, in a letter by him to another of the founders, Karl Marx, dated 12th December 1859:

> Incidentally I am just reading Darwin and find him excellent. One side of theology, the insistence on a personal, transcendent Creator, has not been smashed yet. This is happening now.

For further insight, the Most Rev. Cuthbert M. O'Gara, former Bishop of Yuanling in China, on page 11 of his *The Surrender to Secularism*:

> As R. E. D. Clark shows in his book *Darwin, Before and After*, the "scientific community" accepted Darwinism within a decade of the appearance of *The Origin* (1859). And this was not because of Darwin himself and his book, but because the entire 19th century was marked by the idea of progress and attendant naturalism broadcast by the French Revolution (1789).
>
> Scientists and people generally who thought about these things were prepared and predisposed for evolutionism by the work of men like Buffon and Lamarck in

France; by Hutton, Erasmus Darwin (Charles' grandfather, called by A. G. Tilney the real 'Father of Evolution') and Charles Lyell in England; and by the very powerful philosophical movement generated by Kant and Hegel. Darwin was primarily an *agent* rather than a *cause*. He seemed so scientific with his taste and talent for detail – which he piled up to prove *variation* but *never evolution*.

When the Communist troops overran my diocese (in China), they were followed in very short order by the propaganda corps – the civilian branch of the Red forces... The entire population – everyone, for a week or more – was forced to attend the seminar specified for his or her proper category and there willy-nilly to listen to the official communist line.

Now what, I ask, was the first lesson given to all these people? One might have supposed that this would have been some pearl of wisdom let drop by Marx, Lenin or Stalin. Such, however, was not the case. The very first, the *fundamental* lesson was man's descent from the apes – Darwinism.

As all the foregoing quotations clearly show, the purpose of the hoax, unquestionably a deliberate one, is to destroy man's correct view of God, and ultimately his belief in God.

Chapter 11

Darwin's problems with his own theory.

Darwin and his *On The Origin of Species* have featured often in these pages, as of course they should. For the sake of good order and reasonable completeness, a look at both Darwin and his book must be worth making, and especially at his book, the influence and consequences of which have been vast and continue to be so.

I invite consideration of the following extract from *The Origin*'s chapter 6, titled "Difficulties of the theory". The occasional highlightings in italics are mine.

As introduction to it, I repeat, first of all, what Darwin wrote was quoted in a different context back in chapter 2 of this PART III.

> Long before the reader has arrived at this part of my work, a crowd of difficulties will have occurred to him. *Some of them are so serious that to this day I can hardly reflect on them without being in some degree staggered*; but, to the best of my judgement, the greater number are only apparent, and those that are real are not, I think, fatal to the theory.

This, now, is how Darwin, in his book, proceeded from that introductory paragraph:

> These difficulties and objections may be classed under the following heads.
>
> First, why, if species have descended from other species by fine gradations, do we not everywhere see innumerable transitional forms? Why is not all nature in confusion, instead of the species being, as we see them, well defined?
>
> Secondly, is it possible that an animal having, for instance, the structure and habits of a bat, could have been formed by the modification of some other animal with widely different habits and structure? Can we believe that

natural selection could produce, on the one hand, an organ of trifling importance, such as the tail of a giraffe, which serves as a fly-flapper, and, on the other hand, an organ so wonderful as the eye?

Thirdly, can instincts be acquired and modified through natural selection? What shall we say to the instinct which leads the bee to make cells, and which has actually anticipated the discoveries of profound mathematicians?

Fourthly, how can we account for species, when crossed, being sterile and producing sterile offspring, whereas, when *varieties* of species are crossed, their fertility is unimpaired?

Fifthly, organs of extreme perfection and complication. *To suppose that the eye, with all its inimitable contrivances for adjusting the focus to different distances, for admitting different amounts of light, and for the correction of spherical and chromatic aberration, could have been formed by natural selection, seems, I freely confess, absurd in the highest possible degree.*

Yet reason tells me that, if numerous gradations from a perfect and complex eye to one very imperfect and simple, each gradation useful to its possessor, can be shown to exist; if, further, the eye does vary ever so slightly, and the variations be inherited, which is certainly the case; and if any variation or modification in the organ be useful to an animal under changing conditions of life, then the difficulty of believing that a perfect and complex eye could be formed by natural selection, though insuperable to our imagination, can hardly be considered real. How a nerve comes to be sensitive to light, hardly concerns us more than how life itself first originated; but I may remark that several facts make me suspect that any sensitive nerve may be rendered sensitive to light, and likewise to those courses of vibrations of the air which produce sound...

He who will go thus far, if he find on finishing this treatise that large bodies of facts, otherwise inexplicable, can be explained by the theory of descent, ought not to hesitate to go further, and to admit that a structure even as perfect as the eye of an eagle might be formed by natural selection, even though in this case he does not know any of the transitional grades. His reason ought to conquer his imagination; though I have felt the difficulty far too keenly to be surprised at any degree of hesitation in extending the principle of natural selection to such startling lengths.

If it could be demonstrated that any complex organ existed, which could not possibly have been formed by numerous, successive, slight modifications, my theory would absolutely break down. But I can find out no such case.

In the above, Darwin asks question after question, all of them clearly requiring adequate answers. He ends up even going so far as this: first of all, to admit that to suppose that the eye, with all its "inimitable contrivances" for various purposes, has been formed by natural selection, "seems, I freely confess, absurd in the highest possible degree", and then, immediately after those words, to accept that, in consequence, unless answers that make sense can be given to his questions, his theory "would absolutely break down."

And then? How does he resolve the problems raised by those questions?

Other than by making a few assertions which he makes no attempt to show to be valid, let alone *testably* valid, *he makes no remotely convincing attempt to deal with the objections that he has just raised.*

What this means is that, startlingly, astonishingly, even sensationally, Darwin, in those three paragraphs just quoted, *has demolished his own case*, the case to which his entire book of about five hundred pages is devoted. Good reader, you might like to read those paragraphs again in order to obtain a full grasp of the extent of the devastation they inflict on his theory.

* * * * *

Finally, it is worth noting that, according to a British scientist, L. Merson Davies, on page 7 of in a book by him originally published in 1900, *The Bible and Modern Science*:

> It has been estimated that no fewer than 800 phrases in the subjunctive mood (such as "Let us assume", or "We may well suppose", etc.) are to be found between the covers of Darwin's *Origin of Species* alone.

Oh dear! It is supposed to be *science* that we are talking about; not make-believe.

Chapter 12
Summing up.

At this point a summary of what this writer maintains has been established in these pages may be helpful.

First, life, as we see it all around us, simply *cannot* have come into existence by chance. Furthermore, the fact that many of the laws of nature would otherwise not be laws of nature after all should be enough to rule out the theory of evolution in the mind of any reader.

Secondly, on the impossible hypothesis that life *did* come into existence by chance, it *could not possibly* have developed, also by chance, into the countless different forms of life that exist in the world – that is to say, plant life, germ life, insect life, animal life including fish life, and human life, not to mention the various independent plant, insect and animal species *within* those various forms of life.

Thirdly, the fossil evidence is completely discredited as a source of information relating to evolution in the past. In the first place, the fossil "record" is far too incomplete for such a purpose. In the second place, fossils do not – and, once again, simply *cannot* – arrive in their present form in the way that today's scientists try to lead us to believe that they do.

What about the evidence put forward by Professor Coyne and others to show, they claim, that evolution did in fact take place? That is to say, not only the fossils, but also speciation, vestiges, biogeography, natural selection.

Nothing in those five categories of evidence amounts to any sort of scientific proof of evolution. For instance, the evolutionist S. R. Scadding, who has critically examined vestigial organs as evidence for evolution, said this, in an article by him "Do

'Vestigial Organs' Provide Evidence for Evolution?" (on page 173 of volume 5, May 1981, of the periodical *Evolutionary Theory*):

> **Since it is not possible to unambiguously identify useless structures, and since the structure of the argument used is not scientifically valid, I conclude that "vestigial organs" provide no special evidence for the theory of evolution.**

What about Coyne's argument of biogeography? Let us remind ourselves what he said on this, as summarised in chapter 6 of PART II.

According to him, it is obviously pertinent to ask why a creator would create different animals for different continents while nevertheless also giving them so much in common. No creationists, he says, whether those creationists who believe in the Book of Genesis or any other kind of creationists, have ever given any explanation that makes sense of why it is that different kinds of animals have great similarities in different parts of the world. They are simply reduced to invoking, in Coyne's words, "the inscrutable whims of the creator".

But what is there to explain? Animals which have similarities but live in different parts of the world is no argument at all either for or against evolution. I ask Professor Coyne: why *should* God make them all more different than they are?

* * * * *

Back on page 66, I said that an important thing that I should be doing in this book would be examining with reasonable care the most telling arguments that those who believe in evolution use to justify their beliefs. "Mountains of evidence," there is for evolution, I quoted Professor Coyne as saying on page xiv of his book *Faith versus Fact*, and I said that I should be trying to make sure that no evidence of any significance was left unanswered, at least by implication. I believe that this task has now been completed.

I believe, too, that a conclusion that cannot be reasonably disputed or doubted has emerged as a result. In less restrained prose that has been used so far in these pages, and which is surely appropriate for the purpose of bringing the truth home as clearly as it deserves to be brought home, evolutionary theory can fairly and justly be described as, yes, a load of codswallop.

Is that going too far, or even *very much* too far? Subject to any flaw or flaws in the case presented in this book that any readers may come up with, it seems to me that it is not going even slightly too far, but, rather, is simply a summary in plain terms of yet another example of truth being decided by neither majority vote nor even the all-but unanimous verdict of acknowledged experts on any subject under discussion – an example to add to the example of Professor Dingle's experience of top experts on *his* specialist subject, that of relativity theory, as related on pages 11-20.

Chapter 13
Interestingly, however...

As we have seen, the falsity of all versions of evolution is so clear and obvious that, in the case of some of the evidence against it, even a quite-young child can recognise that falsity without difficulty. Interestingly, however, the chances of this collection of essays resulting in the conversion of its readers who are convinced of evolution when they open it for the first time are, sadly, not very high. It is reasonably safe to say that those who are open-minded as to the reality or otherwise of evolution are likely to have become firm-minded against evolution by now, and equally reasonably safe to say that there will be no shortage of people who will continue to consider themselves to be evolutionists even if they recognise individual arguments against it as valid ones.

Why should this be? – and how can I justify my complete confidence that it is so?

The answer is that, remarkably, the supposed reality of evolution having taken place becomes a virtual dogma for Professor Coyne and all committed evolutionists. Yes, it becomes a position that is unshakable no matter how weighty the evidence against that supposed reality.

Why, we may wonder, is this so? How has it come about that it is so, as it undoubtedly is? That is certainly an interesting psychological conundrum, and perhaps even a fascinating one when one becomes aware of it for the first time. That it *is* so, there is no doubt, however, as is shown by quantities of solid and compelling authoritative evidence given by evolutionists.

Of those who have gone into print on the subject in favour of evolution, there are plenty who have been happy to state that they see problems with the theory. Seldom if ever does the recognition of such problems, however great they may be, result in their conversion away from evolution. Evolutionists they remain, no matter what objections to the case in favour of it they recognise as objections.

As we shall be seeing, part of the cause of their steadfast conviction is the clear fact that the part of the academic world relating to evolution-theory is as ruthlessly governed as Professor Dingle found the part of the academic world relating to relativity-theory to be.

Here, to launch us on our brief study of evolutionist-psychology, are two *pro*-evolution American professors, Jerry Fodor, Professor of Philosophy and Cognitive Science, and Professor Massimo Piatelli-Palmarini, Professor of Cognitive Science, in the introductory chapter of their book *What Darwin Got Wrong* (published by Profile Books, London, 2011):

> This is not a book about God; nor about intelligent design; nor about creationism. Our main contention in what follows will be that there is something wrong – quite possibly fatally wrong – with the theory of natural selection; and we are aware that, even among those who are not quite sure what it is, allegiance to Darwinism has become a litmus for deciding who does, and who does not, hold a "proper scientific" world view. "You must choose between faith in God and faith in Darwin, and if you want to be a secular humanist, you had better choose the latter." So we are told.
>
> We do want, ever so much, to be secular humanists. In fact we both claim to be outright, card-carrying, signed-up, dyed-in-the-wool, no-holds-barred atheists. We therefore seek thoroughly naturalistic explanations of the facts of evolution. It is our assumption that evolution is a mechanical process through and through. This is generally in the spirit of Darwin's approach to the problem of evol-

ution. We are glad to be – to that extent at least – on Darwin's side.

Still, this book is mostly a work of criticism; it is mostly about what we think is wrong with Darwinism. Near the end, we'll make some gestures towards what we believe a viable alternative might be; but they will be pretty vague. In fact we don't know very well how evolution works.

In all informed opinion in fields that either of us has worked in, neo-Darwinism is taken as axiomatic; it goes literally unquestioned. A view that looks to contradict it, either directly or by implication, is *ipso facto* rejected, however plausible it may otherwise seem. Entire departments, journals and research centres work on this principle.

And the chapter continues for several more paragraphs along similar lines. *Nothing* would change their belief in evolution into non-belief, and additional evidence lies in the fact that some of their examples of "what Darwin got wrong" are devastating to evolutionary theory.

In the words of the cliché, you couldn't make it up. No *anti*-evolutionist author would dare to put such thoughts into the minds of evolutionists, let alone represent those thoughts as general ones, indeed even obligatory ones. Look once again at some of what was said by these two professors, this time in summary:

Allegiance to Darwinism has become a litmus for deciding who does, and who does not, hold a 'proper scientific' world view...

In all informed opinion in fields that either of us has worked in, neo-Darwinism is taken as axiomatic; it goes literally unquestioned...

Entire departments, journals and research centres work on this principle.

Our next witness is a former Darwinist who changed his mind and became a genuine anti-Darwinist, Professor David Gelernter, a Yale University professor and a deservedly famous one, admired for having predicted the World Wide Web several years before it came into existence, and the developer of a considerable number of computing tools. Rather than quoting directly what he says, I am taking the following from an article about him by an author, Jennifer Kabanny, who published the following in *The College Fix* – a website devoted to higher education – which she reinforced by quoting him from time to time.

https://catholiccitizens.org/uncategorized/88405/famed-yale-computer-science-professor-quits-believing-darwins-theories/

In May, the *Claremont Review of Books* published a column by Gelernter headlined "Giving Up Darwin." In it, Gelernter explained how his readings and discussions of Darwinian evolution and its competing theories, namely intelligent design, have convinced him Darwin had it wrong. The professor expanded on his views in an interview with Stanford University's Hoover Institution that was published last week.

Gelernter stops short of fully embracing intelligent design, both in his essay and during his interview. "My argument is with people who dismiss intelligent design without considering, it seems to me — it's widely dismissed in my world of academia as some sort of theological put-up job — that it's an absolutely serious scientific argument," Gelernter said during his interview. "In fact it's the first and most obvious and intuitive one that comes to mind. It's got to be dealt with intellectually."

Gelernter said an ideological bent has taken over the field of science. He said that he likes many of his colleagues at Yale, and that they are his friends, but that when he looks at "their intellectual behavior, what they have published — and much more importantly what they tell their students — Darwinism has indeed passed beyond a scientific argument as far as they are concerned. You take your life in your hands to challenge it intellectually. They will destroy you if you challenge it."

"Now, I haven't been destroyed, I am not a biologist, and I don't claim to be an authority on this topic," Gelernter added, "but what I have seen in their behavior intellectually and at colleges across the West is nothing approaching free speech on this topic. It's a bitter, fundamental, angry, outraged rejection [of intelligent design], which comes nowhere near scientific or intellectual discussion. I've seen that happen again and again."

Gelernter acknowledges that he is, in his words, "attacking their religion," and graciously adds: "And I don't blame them for being all head up. It is a big issue for them."

How does the field of biology get over Darwin? Gelernter said the outlook is bleak. "Religion is imparted, more than anything else, by the parents to the children," he said. "And young people are brought up as little Darwinists. Kids I see running around New Haven are all Darwinists. The students in my class, they're all Darwinists. I am not hopeful."

Note, in the above, the following in particular:

"An ideological bent has taken over the field of science...

"Darwinism has indeed passed beyond a scientific argument as far as his colleagues at Yale are concerned. You take your life in your hands to challenge it intellectually. They will destroy you if you challenge it...

"What I have seen in their behavior intellectually and at colleges across the West is a bitter, fundamental, angry, outraged rejection [of intelligent design], which comes nowhere near scientific or intellectual discussion."

"I haven't been destroyed."

Why has he not been destroyed? The *only* reason, he makes it clear, is that he is not in the same department of Yale University as his fellow-Yale-academics are.

Probably those three witnesses are sufficient for our purposes, but so important is it to be fully aware of the principal reason for the widespread belief in evolution notwithstanding the lack of evidence in its favour and the quantity of evidence in opposition to it that I think it as well to risk overdoing the exposure of it than to take even the smallest risk of not giving fully sufficient evidence. I call to witness finally, therefore, the author and journalist Michael Brooks PhD, a former editor of the *New Scientist* magazine. The following colourful and extremely revealing passages are from his book *Free Radicals: The Secret Anarchy of Science* (Overlook Press, New York, 2011), written with reference to today's scientific world in general rather than specifically to the evolution area of science.

> **Science is peppered with successes that defy rational explanation, and failures that seem even more illogical... This is not the "wacky" science, the crazy things that happen on the fringes of research. This is the *mainstream*. These anarchists are behind many of the Nobel Prizes of the last few decades... It really does seem that, in science, anything goes. And this is no new phenomenon. *Science has always been this way*.**

"For over half a century," Brooks continues:

> ...scientists have been involved in a cover-up [about how science actually works] that is arguably one of the most successful of modern times....

Everything in science, he says, has been affected, including

> ...the way it is done, the way we teach it, the way we fund it, its presentation in the media, the way its quality-control-structures – in particular, peer review – work (or don't work), the expectation we have of science's impact

on society, and the way the public engages with science (and scientists with the public) and regards the pronouncements of scientists as authoritative. We have been engaging in a caricature of science, not the real thing.

Brooks devotes more than three hundred pages to documenting those claims, and this is how he summarises the conclusions that, he maintains, necessarily follow from what he has shown:

Science is the fight to the intellectual death, but not between equal adversaries. It takes place in a gladiatorial arena where the challenger has to overcome, not only the established champion, but also his or (more rarely) her supporters. And, whether in attack or defence, the fight is rarely clean.

Thus my grounds for thinking it reasonable to suppose that few if any firm believers in evolution will change their positions after reading this book even though – we can be confident – they will not be able to come up with any arguments of any genuine significance against the conclusions drawn in it.

Readers may perhaps recall that back in the first chapter of this book I said this, after recounting what Professor Dingle went through when he found that Einstein's theory of relativity was fatally flawed:

"To any reader who may think that this picture of the world of science exists only in the arena of relativity, I make the suggestion that we should keep in our mind at least the *possibility* that it may apply in other areas of science as well."

Do my gracious readers now agree with me that this was sound advice?

It seems to me fitting to close by repeating what I said very near the beginning of this book, specifically on page 4.

Sadly, there can be no doubt that all too many of those who are confronted with what is in these pages will be *primarily* affected in their judgement, *not* by the *reasoning and evidence* supporting one side or the other, as of course they ought to be, but, rather, by their *present beliefs or convictions*, in effect by their mental *habits*.

Deep-set mental habits, what is more. Deep-set mental habits that are conditioned from the moment our education starts and from then on throughout our lives – by modern textbooks on science and on history, and from what we are told in newspapers and on radio and television by astronomers, geologists and other "evolutionist" scientists, and, for that matter, by virtually everyone else who makes any sort of public statements on such matters.

For whatever reason, just as it is safe – or at least I *hope* that it is safe – to assume that, after reading this book, anti-evolutionists are likely to remain anti-evolutionists, it is *also* the case that evolutionists will tend to want to remain evolutionists.

Please, good reader, do not be one of the countless people who are "dictated to" by their mental habits. Rather, make a firm resolve to be guided by reason and adequate evidence, and *only* by reason and adequate evidence, and by nothing else.

* * * * *

Because of the importance – it would indeed be difficult to think of any subject that is *more* important – of what has been covered in this book, I consider it appropriate to invite readers to make contact with me by e-mail if they have any questions to raise, objections to make, or any other reason to wish to do so.

E-mail address: nmg@nmgwynne.com

Chapter 14

Post Script.

Some scientists on science as it is today.

In the chapter labelled Introductory, at the beginning of this book, and again in the previous chapter of this book, chapter 13 of PART III, I devoted some space to showing from authoritative sources that modern science is not to be trusted in any of the fields that it is needed for. I believe I can reasonably suppose the evidence that I gave in support of this proposition to be sufficient for all practical purposes; but so important is this reality about science, because of how completely science dominates so many subjects, some of them of the highest importance, in today's world, that I believe it worth adding further evidence when such evidence is particularly telling.

A significant source of information on this aspect of science is an exceptionally scholarly book published as recently as in 2017, *Evolution's Blunders, Frauds and Forgeries* – a title that is fully supported by the book's contents – by Jerry Bergman Ph.D. The following information is taken from its penultimate chapter, Chapter 18. (Each paragraph is from a different part of the chapter.)

> Fraud as a whole in science now is "a serious, deeply rooted problem" that affects no small number of contemporary scientific studies, especially in the field of evolution. (M. Roman in an article "When good scientists turn bad", *Discover* 9(4), 1958: page 58.) Scientists recently have been forced by several events to recognise this problem and to attempt to deal with it....

*   *   *   *   *

> The fraud problem is worldwide, and has been reported in the United States, Britain, France, Germany, India,

Japan, Poland, Sweden and Australia... (H. F. Judson: *The Great Betrayal Fraud in Science*. (Harcourt, New York, 2002: page 132.)

* * * * *

Fraud does not just exist, but is epidemic to the extent that one study about the problem concluded that "science bears little resemblance to the conditional portrait."... (*Nature*, 415 (6872), 2002: page 597.)

* * * * *

The present system of science actually encourages deceit. Careers are at stake, as all jobs, grants, tenure, and, literally, one's livelihood. (Dalton, Peers under *Pressure*, *Nature*, 413, 2002 (6852: page 104.) This problem is expressed in the saying that a professor must publish or perish (meaning that he will be denied tenure otherwise, which often ends an academic career. As Broad and Wade note on page 36 of their book *Betrayers of the Truth: Fraud and Deceit in the Halls of Science* (Simon & Schuster, New York, 2002), "Grants and contracts from the Federal Government... dry up quickly unless evidence of immediate and continuing success is forthcoming." The motivation to publish, to make a name for oneself, to secure prestigious prizes, or to be asked to join a scientific society, all entice dishonesty.

* * * * *

Broad and Wade's disturbing conclusion is that corruption and deceit are just as common in science as in any other human undertaking, because "scientists are not different from other people. In donning the white coat at the laboratory door, they do not step aside from the passions, ambitions and failings that animate those in other walks of life." (Ibid.: page 19.)

* * * * *

The exaggerated claims about peer review ensuring quality science are a myth. The former editor-in-chief of

Science, Donald Kennedy, admitted that "peer review has never been expected to detect scientific fraud." (A. Anderson: "Conduct unbecoming: it's the biggest scandal ever to hit physics" in New Scientist 176, 2363, 2002: page 3.)

*   *   *   *   *

The reasons for this serious problem in science include the drive for money, tenure, promotions, grant renewals, professional rivalry, and the need to prove one's own theories and ideas. Another factor is the rejection of Christianity and moral absolutes, which has resulted in a collapse of the moral foundation that is critical in controlling fraud. Some people claim that the problem is so serious that possibly the majority of published research claims are false. (J. P. A. Ionnidis, "Why most published research findings are false." *PLoS Medicine*, 2(8) 2005: pages 696-701.)

Bibliographical.

A large, indeed very large, number of books, booklets, etc., has been written promoting one side or the other of the topic covered in this book since the time that it was basically started up by Charles Darwin in the mid-nineteenth century. No useful purpose would be served by an attempt to include all of them here, and I shall therefore attempt only a reasonably representative selection that includes all the books that I have quoted from in the preceding pages.

Promoting or tending to promote at least not opposing the theory of evolution.

Coyne, Professor Jerry A., Professor Emeritus in the Department of Ecology and Evolution in Chicago University: *Why Evolution is True* (Oxford University Press, 2009) and *Faith Versus Fact: Why Science and Religion Are Incompatible* (Penguin Books, U.S.A, 2015)

Cronin, R.: *The Ant and the Peacock: Altruism and Sexual Selection from Darwin to Today* (Cambridge University Press, 1991)

Darwin, Charles: *On the Origin of Species by Means of Natural Selection, or the Preservation of Favoured Races in the Struggle for Life* (John Murray, London, 1859)

Dunbar, R. L. Barrett: *Evolutionary Psychology: A Beginners Guide* (One World, Oxford, 2005).

Forrest, P., and P. R. Gross: *Creationism's Trojan Horse: The Wedge of Intelligent Design* (Oxford University Press, 2007).

Futuyma, D. J.: *Science on Trial – The Case for Evolution* (Sinauer Associates, Sunderland, Massachusetts, 1995).

Gibbons, A.: *The First Human: The Race to Discover Our Earliest Ancestors* (Doubleday, New York, 2006).

Hawking, Professor Stephen, and Doctor Leonard Mlodinow: *The Grand Design* (Bantam Books 2010).

Himmelfarb, Gertrude: *Darwin and the Darwinian Revolution* (Chatto & Windus, London, 1959). It could in fact be argued that this book belongs to the next section, because the author really does try to give all relevant information and does not hesitate to criticise some of the advocates of Darwinism. Indeed, it is even sometimes difficult to work out which side she is on.

Johanson, D., and B. Edgar: *From Lucy to Language* (revised edition – Simon and Schuster, New York, 2006).

Kitcher, P. J.: *Living with Darwin – Evolution, Design, and the Future of Faith* (Oxford University Press, New York, 2006).

Mayr, B.: *What Evolution Is* (Basic Books, New York, 2002).

Mindell, D.: *The Evolving World – Evolution in Everyday Life* (Harvard University Press, Cambridge, Massachusetts, 2007).

Prothero, D. R.: Evolution: *What the Fossils Say And Why It Matters* (Columbia University Press, New York, 2007).

Ruse, Michael, and Joseph Travis (editors): *Evolution: The First Four Billion Years* (Harvard University Press in 2009).

Scott, E. C.: *Evolution Versus Creationism – An Introduction* (University of California Press, Berkeley, California, 2005).

Wysong, Dr. R. L.: *The Creation Evolution Controversy* (Inquiry Press, Michigan, 1976)

Opposing the theory of evolution, or at least tending to oppose it.

Andrews, Professor E. H.: *From Nothing to Nature* (Evangelical Press, U.K., 2000)

Barnes, Thomas G.: *Origin and Destiny of the Earth's Magnetic Field* (1971)

Bergman, Jerry: *Evolution's Blunders, Frauds and Forgeries* (Creation Book Publishers, 2017)

Bowden, M.: *The Rise of the Evolution Fraud* (Second edition: Sovereign Publications, Kent, 2008)

Clark, R. E. D.: *Darwin, Before and After – The Story of Evolution.* (Paternoster Press, London, 1948)

Courville, Donovan A.: *The Exodus Problem and its Ramifications* (Crest-Challenge Books, Ca., 1971)

Custance, Arthur C.: *Genesis and Early Man* (Zondervan of HarperCollins Group, 1975)

Davies, L. M.: *The Bible and Modern Science* (1925)

de Vries, Hugo: *Species and Varieties* (Chicago, 1905)

Dewar, Douglas: *Challenge to Evolutionists* (Uplift Books, third edition 1948)

Fides, Anthony Michael: *Man's Origins* (Augustine Publishing Company, Devonshire, 1977)

Field, A. N.: *The Evolution Hoax Exposed* (Reprinted Tan Books, North Carolina, 1971)

Fodor, Jerry, and Massimo Piatelli-Palmarini: *What Darwin Got Wrong* (Profile Books, London, 2011)

Gallop, Roger: *Evolution: The Greatest Deception in Modern History* (Red Butte Press, inc., Florida, 2011; second edition 2014).

Hiebert, Henry: *Evolution: Its Collapse in View* (1979)

Johnson, Wallace: *The Case Against Evolution* (Catholic Centre for Creation Research, 1976)

John Kasper, John: *Gifts from Agassiz: Or Passages on the Intelligence Working in Nature* (Literary Licensing, Montana, 2013)

Lewis, C. S.: *Abolition of Man* (Oxford University Press, 1943)

Lewis, C. S.: *Mere Christianity* (Geoffrey Bles, 1952)

Morris, Henry M.: *Many Infallible Proofs – Practical and Useful Evidences of Christianity* (Master Books, Arizona, 1996)

Morris, Henry M.: *The Bible and Modern Science* (Moody Press, Chicago, 1951)

O'Gara, The Most Rev. Cuthbert M.: *The Surrender to Secularism* (Cardinal Mindszenty Foundation, Inc., St. Louis, Missouri, 1967)

Nelson, Byron: *After Its Kind* (Bethany House Publishing Group, Bloomington, Minnesota, 1967; originally published in the 1920s)

Patten, Donald W.: *The Biblical Flood and the Ice Epoch* (*Pacific Meridian Publishing Co, Seattle, WA.* 1966)

Schumacher, E. F.: *A Guide for the Perplexed* (Vintage of Penguin Books, London, 1997)

Sheehan, Rev. Michael: *Apologetics and Catholic Doctrine* (M. H. Gill & Son, Dublin, 1950).

Stove, David: *Darwinian Fairytales – Selfish Genes, Errors of Heredity, and Other Fables of Evolution* (Encounter Books, New York, 1995)

Watson, D. C. C.: *Myths and Miracles* (Publisher unknown 1976)

Watson, David C. C.: *The Great Brain Robbery* (self-published by the author, 5th edition 1989)

Whitcomb, John C., and Henry M. Morris: *The Genesis Flood* (originally published, 1962)

Relevant to the subject matter of this book but not expressly addressed to its topic.

Dingle, Herbert: *Relativity for All* (1922; an essay reprinted by Hardpress Publishing, USA)

Dingle, Herbert: *The Special Theory of Relativity* (a monograph published in 1940)

Gillis, Rev. James M.: *False Prophets* (1929; republished Kessinger Publishing, 2010)

Knox, Monsignor Ronald: *The Beginning and the End of Man* (publisher and date of publication unknown)

Morris, Henry M.: *Many Infallible Proofs – Practical and Useful Evidences of Christianity* (Master Books, Master Books, Arizona, 1975)

Morris, Henry M.: *The Remarkable Birth of Planet Earth* (Dimension Books 1972)

Acknowledgements.

I should like to put on record my gratitude to the following people, for their help in some cases with parts of this book and in some cases with the whole of it.

Very much first, in order of those to whom I am indebted, is undoubtedly Mr. Hugh Williams of St. Edward's Press, this book's publisher. Right from the outset of its being agreed between us that it should be written, with St. Edward's Press its publisher, and finding himself having to read several drafts before it reached finalisation, his constant enthusiasm and his attention to the minutest detail made an immeasurable contribution to the book as it has turned out to be.

Second in importance: my daughter and my fellow-teacher in our Gwynne Teaching organisation, and also the administrator of it, Miss Chloe Gwynne, who reorganised PART III extensively and made other valuable contributions to this second edition of this book.

The following people have been kind enough to offer their comments, and between them have made many valuable contributions, sometimes important ones: His Royal Highness the Duke of Kent; Charles Beauclerk, Earl of Burford; and – from now on in alphabetical order rather than in order of traditional social precedence! – Mr. David Castelnovo, Mrs. N. M. (Frederica) Gwynne, Miss Patricia Horn, Mrs. Derek (Stephanie) Kenna, Mr. S. B. Kumaramangalam, Mrs. Sean (Anne) Murphy, Mr. Robert Pols, Mrs. Michael (Liz) Spicer and Miss Sophia Talbot Rice.